Merchant of the Medomak

Stories From Waldoboro Maine's Golden Years

1860 – 1910

by

Mark W. Biscoe Mark W. Biscoe

Schooner Dorothy Palmer *launched June 4, 1903. Author's collection*

Cover picture: *Down Easter* Emily Reed - *built by A.R. Reed, 1880*

First edition
September 2004

Printed in the United States of America
by Lincoln County Publishing Co., Newcastle, Maine 04553

ISBN 0-9652746-1-6

Table of Contents

With Deepest Gratitude

The publishing of this volume would not have been possible without the generous support of the following men and women:

William Achtmeyer
John and Marilyn Andrews
Reed and Barbara Anthony
Thomas and Barbara Beal
Mark and Therese Biscoe
William and Carol Blodgett
Pierce B. Browne
Carolyn Bryant
Theodore Clapp
Samuel and Harriet Cohen
Christopher and Susan Davies
John and Jane Farlow
Frances (Kitty) Fassett
Kay Flanagan
Marjorie Freeman
Stephen and Dorothy Gardner
Robert and Karen George
Emory A. Hamilton
Thomas and Anne Jordan
Barbara Keyes
Lawrence, Michelle, and Ethan Lasser
Peter and Judith Lawrence
Patricia Magee
Henry and Mary Myers
George and Carole Petrowsky
Thatcher Pinkham, Jr.
J. Rivington and Jane Surran Pyne
Frank and Carolyn Rasmussen
Elizabeth Sproul
Alfred and Mary Storer
Maryanne P. Storer
Arthur and Eleanor Travers
Robert and Virginia Travers
Ewing and Dona Williams
Charles and Cynthia Wright

The Waldoborough Historical Society
The Waldoboro Woman's Club

Acknowledgements

A research project on local history makes the researcher a firm believer in the old adage: "*No man is an island*." Without the friendly and freely given aid and advice from the following persons and institutions, this project would never have been completed. My thanks and appreciation go to them.

The staffs of the Waldoboro, Rockland, Skidompha (Damariscotta), and Bangor Public Libraries, The Penobscot Marine Museum, The Maine Maritime Museum, the Lincoln County Court House (Wiscasset), the Waldoborough Historical Society, the Rockland Historical Society, and the Peabody – Essex Museum (Salem, Massachusetts).

The writings of historian and newspaper editor Samuel L. Miller. The writings of historian Jasper J. Stahl. The ship lists compiled by these two gentlemen and augmented by the research of customs house agent and marine historian, Robert Applebee as well as customs house agent Lowell Sidelinger. The imagination and artistry of Tom Turlo.

Elsie Eugley, Martha Boggs, Darryl McKinney, Alfred (Fred) Storer, William Blodgett, Richard Wallace, Bill and Jean Travers, Dr. David Smith, Dr. George Dow, Kay Flanagan, Charles Begley, Alan Benner, Van Reid, Randy Gross, Thomas Jordan, Steve Cartwright, Robert Pearson, Dr. Charles Burden, Susan Porter, Mike Footer, Philip Achorn, Delma Baldwin, George Leach, Sandra Denman, John Gatcombe, J. Malcolm Barter, Marjorie Freeman, Louis Boissoneault, Ted Wooster, Nick Dean, Elizabeth Sproul, Gordon Webster Sr., Gordon Webster Jr., Trudy Chambers Price, Tom Moody, John Murray Jr., Captains Douglas and Linda Lee, Nathan Lipfert, Captain W.J. Parker, George Wright, the late James Stevens, and the late Stanley Bailey.

Particular thanks goes to a true Waldoborian, Stan Hanna, who, at ninety years of age and more, shows a unique example to all of us for vigor, feistiness, humor, courage, and an unfailing memory about and interest in his home town.

And, as ever, my gratitude and love to my wife Jane – proofreader, advisor, and calming presence extraordinaire – who has put up with my peculiar ways of starting, stopping, and becoming hopelessly ensnared in projects (whether they be basketball teams or research) for 45 wonderful years.

Foreword

(Author's note: Dr. David C. Smith, contributor of the following remarks, served as my faculty advisor in the University of Maine graduate program from 1963 to 1967. It has been my distinct privilege to know him during and following my earning a Master's Degree in History. To my wife and me, it seemed only appropriate to approach Dr. Smith for assistance in the writing of this book; for he had been such a helpful friend 40 years ago.)

"The profession of teaching has been an integral part of my life – laying out a syllabus, organizing the research which had been less than perfect the last time I did the course, and actually talking to the class once the semester came around again. All the years I went into the classroom (and still do, when invited) I always look forward to the world of the teacher and the student.

There are some subsets of this work, however. A great thrill one can have is when a student comes into the office and says, 'Have you read this?' or 'Here is that book you mentioned.' This means that the teacher and the student have begun that dialogue which is so much a part of the profession of teaching. From my experience as a faculty member at the University of Maine in Orono and Farmington, Cornell University, and William Smith College, there are still students with whom I correspond by mail, E-Mail, telephone, and by personal visits. We teach each other.

Probably the greatest aspect of such communications is when a student from the past shows you his **New Book** and says, 'Will you read…?' The book you now hold in your hands is just such an example. Mark Biscoe, former and still student of mine, and now a teacher of mine as well, has spent a great deal of time and effort in collecting, organizing, and stitching together (much like an old time sailmaker) the material that results in *Merchant of the Medomak*, a chronicle of a maritime village in the late 19th century.

This book gives us a close view of the lives of shipbuilders, officers, crews, shipping agents, families, neighbors, and friends living and working in Waldoboro, Maine. The first chapter recounts stories of the ships and men – the Storers, Nashes, Kennedys, Reeds, Welts, Fishes and other who made fortunes from constructing the vessels. The second chapter describes some of the adventures – often very far from home – that the vessels and men encountered. The third chapter describes life in Waldoboro village, and finally, the book turns to the last hurrah – 1900-1904 – when William Palmer, Boston businessman, and the ambitious local citizen, George Welt, brought together their talents and often grating personalities to create a massive fleet of schooners.

The story of these people, their vessels and this town is what good history is all about. Thanks to Mark (and his right hand person, Jane!) 'Here is another book, Dr. Smith.' Thank you indeed."

David C. Smith
Professor Emeritus
University of Maine
August 2004

Dedication

To our children – Mark, Andy, and Kate – and their families – with our love

and

To the people of Waldoboro, Maine – with the hope that this book will remind them of the glorious past of their town as they build for a strong future.

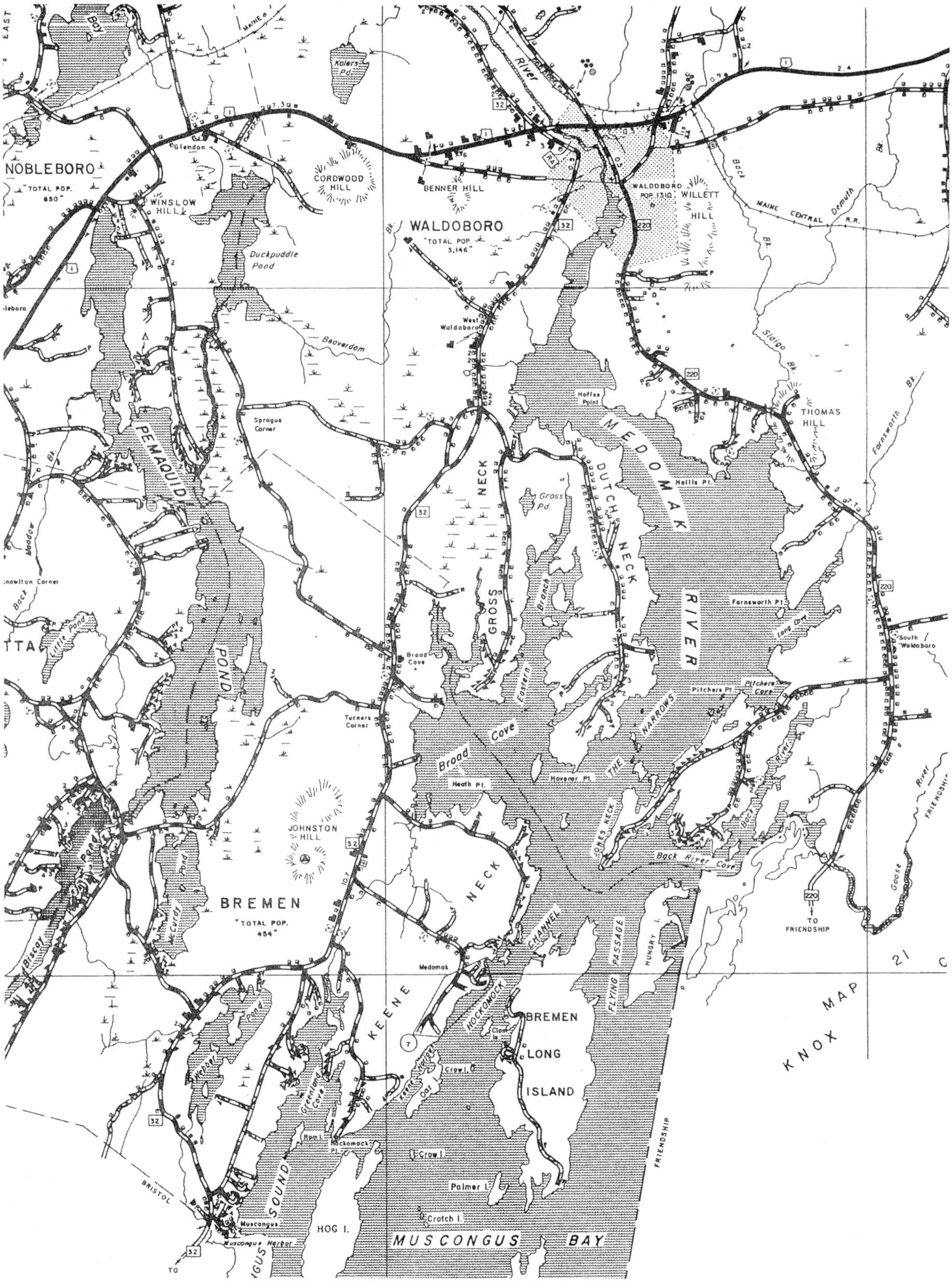

Map courtesy of the Waldoboro Town Office

List of Photographs and Maps

Introduction – Waldoboro, Maine

Waldoboro was a major New England shipbuilding and shipping center from the late 1700's until the early 1900's. Although, as in many Maine communities, records are scarce, we can be sure that over 600 sailing vessels were built, mainly within one half a mile of the village. The map on page 8 indicates approximately where the various shipyards were located. (There are many yards we do not know as yet. Readers must understand that vessels were often built on the outskirts of an individual's property, and many yards were used over and over by different craftsmen.)

The most common type of vessel built in Waldoboro was the schooner (see page 3) – two or three masts – a "coaster" – for bulk cargoes such as wood – to be exported, in the early days from local parts to the large cities of the East Coast. From 1820 to 1840, for example, there were 12 schooners, 11 of which were built locally, which ran regularly between Waldoboro and Boston. Schooners operated like the trailer trucks of today – traveling back and forth, up and down the coast, and making many trips each year. Cargo of any kind – including animals and people – was their business.

After about 1850, much of the fine wood for shipbuilding had been depleted along the Maine coast and the schooners – now larger and with more masts – began to be used in importing wood for the yards. The May 15, 1874 *The Lincoln County News* reported that 11 cargoes of lumber had arrived that week from southern states. Several of the vessels had to wait for days before being able to unload at the Waldoboro wharves and shipyards.

Other types of vessels built in Waldoboro were brigs, barks, barkentines, ships, brigantines, and sloops. (See page 3.) Many of the largest of these, particularly the ships, were sold to prominent firms that used them in the trade to such faraway places as Europe, Asia, Africa, and Australia. And, as the 19th century reached the second half, larger and larger craft were being produced. In 1841 Joseph Clark (see shipyard map) built *Desdemona,* 625 tons, which was the largest craft until he built *Caroline and Mary Clark,* 800 tons in 1849. Then, in 1854, B.B. Haskell & Co. launched the ship *E. Wilder Farley* of 1300 tons. Clark's ship *Joseph Clark* in 1856 was measured at 1308 tons, but this figure was outstripped in the same year by Alfred Storer and Charles Comery's *Hamilton Fish* of 1628 tons. Edwin O. Clark, Joseph's son, built the ship *Mabel Clark* of 1661 tons in 1877; A.R. Reed completed the *George Curtis* – 1745 tons in 1884, and she remained the largest of all until the famous Palmer schooner fleet of George Welt. These six were the *Fannie Palmer,* 2075 tons in 1900; *Baker Palmer,* 2240 tons in 1901; *Paul Palmer,* 1763 tons in 1902; *Dorothy Palmer,* 2315 tons in 1903; *Singleton Palmer,* 2357 tons in 1904 and finally *Harwood Palmer,* 2400 tons in 1904.

The peak years for Waldoboro shipbuilding were from 1840 to 1860 when 225 vessels were built. At that time, because of this industry, the town enjoyed its zenith in population – 4569. The following breakdown in the population figures seems to be directly linked to the fortunes in shipbuilding:

Year	1790	Population	1205
	1800		1500
	1820		2449
	1840		3661
	1860		4569
	1870		4174
	1880		3758
	1890		3506
	1900		3145

Our records show that the most ships built in one year were: 22 in 1848; 23 in 1849; and 21 in 1850. In that period, those 66 ships were built by the following firms: Thomas Gay, Reed & Welt, J.R. Groton, Wm. Fish, Vannah & Hall, Henry Kennedy, Genthner & Morse, Harriman, Samuel Nash, M.M. Rawson, Robert Miller, William Welt, Jacob Hahn, Joseph Clark, Edwin Achorn, George Sproul, Anthony Castner, Fred Creamer, John Levansaler, and Alfred Storer.

By 1853 Waldoboro had become the name for a customs district stretching from Bristol to Thomaston and producing 23% of the total tonnage from the state of Maine (number one of all states in shipbuilding). This 23% was second only to the Bath customs district and far more than Portland, Belfast and Wiscasset combined! Waldoboro village was a busy, thriving place with many stores, hotels, large homes, and a bustling waterfront. Vessels appeared from down river on a daily basis to drop off and pick up cargoes. Sounds from the shipyards – the pounding hammers, whining saws, and plinking caulking irons could be clearly heard by the many people – visitors and shoppers – filling the wooden sidewalks and gravel streets. Prosperity was to be had for many who invested their money in local ships, often paying their investors back in the very first voyage!

This prosperity was to continue through the 1850's, suffer somewhat in the '60's due to the Civil War, and then gradually decline and die out in the 70's, 80's and 90's. Only the wealthier yards of Storer, Kennedy, Fish, Clark, and Reed/Welt could afford to battle against the ever improving steam and steel ships built elsewhere. Toward the end of the century Waldoboro shipbuilder Levitt Storer built the first 5 masted schooner ever constructed in the U.S. – the famous *Gov. Ames,* and in the early 20th century George Welt gave more hope that the industry was to be revived with his six giant Palmer vessels; but it was not to be – the forces bringing wooden shipbuilding and Waldoboro's glory days to an end were firmly in control. Prosperity was at an end, sadly.

But it is wonderful to look back upon...

Chapter 1
The Shipbuilders

Schooner Clara G. Loud *leaving Waldoboro – 1872*
Looking West from the Clark shipyard. Note the large piles of cordwood ready to be shipped.
From the collections of Captains Douglas K. and Linda J. Lee and W. J. Lewis Parker

Typical Waldoboro Vessel – Two Masted Schooner

Irene Meservey – *114 X 29 X 10 feet; John Wall, Master; Owned by Irene Meservey et al – 1866 1870;* Albion Meservey – *1872 – 1882; Bangor Lumber Company – 1917 – 1925; abandoned on the Brewer flats 1925. From the files of marine historian Robert Applebee.*

Vessel Rigs and Numbers

It has been extremely difficult to calculate the exact number of sailing vessels built in Waldoboro. Destructive fires and other acts of nature plus careless handling of shipping records make lists of ships, their tonnages, their names, their rigs, and their builders, full of question marks. However, it is possible to make several interesting assumptions about Waldoboro's great industry. One is that there were approximately 600 sailing ships built in Waldoboro between 1783 and 1904. Another is that a majority of these vessels was the schooner – a two or three masted craft built for coastal commerce – 80 to 150 feet long and capable of carrying 100 to 250 tons of cargo. A third is that the really large vessels – full rigged ships and four or five masted schooners – 200 to 300 feet long and carriers of 700 to 2000 + tons – seldom if ever found their way back to their home port – they were simply too large for the Medomak River! Most of these huge vessels were built by the same firms – Storers, Reeds, Welts, Achorns and Clarks – and in the time periods of the 1850's and 1860's, the 1880's, and 1900–04.

Types of vessels built in Waldoboro and the approximate numbers from 1783 (the first recorded) to 1904 (the last recorded):

Schooners – 1783 to 1854 – 193
1854 to 1904 – 98

Three masted schooner

Brigs – 1783 – 1854 – 90
1854 – 1904 – 30

Brig

Barks – 1783 – 1854 – 45
1854 – 1904 – 16

Bark

Barkentines – 1783 – 1854 – 2
1854 – 1904 – 8

Barkentine

Sloops – 1783 – 1854 – 22
1854 – 1904 – 10

Sloop

Full Rigged Ships – 1783 – 1854 – 41
1854 – 1904 – 35

Full Rigged Ship

Silhouettes of vessels taken from the pamphlet "Rigs of the Nine Principal Types of Sailing Vessels", Charles Davis.

1856 Statistics

Total Tonnage of Vessels Registered, Enrolled and Licensed in Top U.S. Ports:

New York	1,328,000
Boston	521,000
Philadelphia	197,000
Bath	193,000
Baltimore	183,000
New Orleans	163,000
Waldoboro	155,000

Total Tonnage of New Vessels Built, by States:

Maine	149,000
Massachusetts	80,000
New York	76,000
Maryland	19,000
New Hampshire	10,000
New Jersey	9,000
Connecticut	7,000

Total Tonnage of New Vessels Built, by Customs District:

Boston	62,000
Bath	50,000
New York	49,000
Waldoboro	30,000
Portland	21,000
Philadelphia	19,000
Baltimore	15,000
Belfast	11,000
Newburyport	7,000

Compiled from annual *Report on Commerce and Navigation*

Courtesy of Maine Maritime Museum

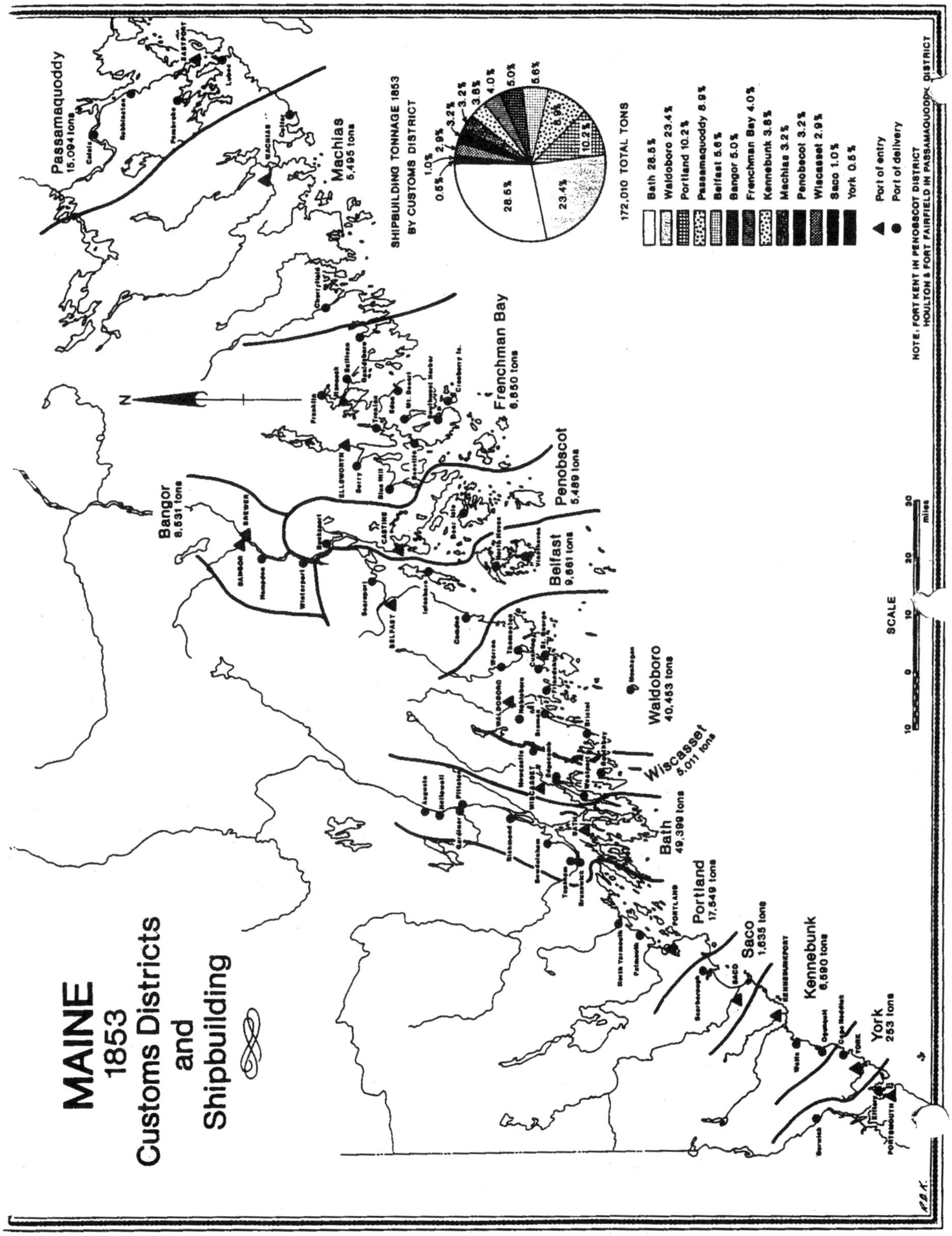

Courtesy Maine Maritime Museum

The Busy Shipyards – 1874

The March 6, 1874 edition of *The Lincoln County News* (a brand new Waldoboro weekly) gushed optimistically over the shipbuilding prospects for the season. "The vessels now upon the stocks in Waldoboro, and the frames that are coming for vessels to be built in the coming summer, amount to 2 ships, 1 barkentine, 10 schooners, and 1 scow. The prospects for a busy season look encouraging."

In May there came a similar comment: "Since '54 our river, wharves, and shipyards have not received the brisk business-like appearance that they do now. No less than 11 cargoes of lumber have been received, several vessels having to wait several days for a berth at the wharf to discharge. The covers are lined with rafts of hard pine, and the yards are piled with other materials to be used in the construction of ships."

Five major shipbuilding firms monopolized the village waterfront in 1874: on the East side of the Medomak were the yards of the first cousins Joseph Clark and Henry Kennedy; on the West side were William Fish; Isaac Reed and Augustus Welt; A.R. Reed, and the Storer father and son, Alfred and Levitt.

The Clark yard, operated by Joseph and later by his son Edwin, lay just south of the bridge. From 1835 to 1883 these men built more than 40 vessels – including two in 1874 – the 500 ton barkentine *Josephine* launched in early August after having been started in February and the 1300 ton ship *Carrie Clark*. Two newspaper reports indicate that dynamite was used liberally by the Clarks to blast the ledge at the stern of one of their vessels in the launching process. These heavy explosions were "startling to the neighborhood."

Henry Kennedy & Company was located directly south of the Friendship Street brick schoolhouse and was responsible for about 30 vessels built between 1837 and 1880. By October 1874 the schooner *Mary J. Cook* had been launched and lay embedded in the mud, as was the common custom, for several weeks, until she could be hauled off. By October 30 Kennedy had stretched the keel of another schooner, to be named *Florence L. Genovar* and launched in 1875.

On the West side of the river, just south of the bridge, energetic businessman William Fish built from 1860 to 1883 a fleet of approximately 20 smaller vessels – mostly schooners. Eighteen seventy-four saw Fish beginning on his largest – a barkentine of 469 tons. Proudly the *News* announced her launch on November 27:

"Launched Monday by William Fish the barkentine *Fred Eugene* – 469 tons – A – 1 for 12 years – length of keel 138', overall 142'; frame of white oak; hardwood floor planking 4" thick; 3 keelsons 14" X 14"; sister keelsons iron fastened through and through; lower deck beams 12" X 14" hard pine; decks are white pine 3" thick; 1500 locust treenails wedged. Workmen included: Master Workman Luvis Benner; Joiners Reizer and Waltz; Blacksmith Andrew Sides; Painter J.W. Miller; Caulker Miles Standish; Rigger John Longley; Sailmaker Charles Redman. The vessel was owned by the builder William Fish, Esq., her commander Captain Adam Smalley, and parties in Portland. Named for one of the partners in the house of George S. Hunt, Portland, the *Fred Eugene* will be a credit to him and to Waldoboro."

A few hundred yards south of William Fish's establishment, approximately where the Town Landing is today, was the yard of Isaac Reed the younger and Augustus Welt, and in a separate business, Isaac's son Asa Redington (or A.R.) Reed. From 1837 to 1904 these men, or other members of their families, built more than 50 vessels, with at least 12 measuring over 1000 tons. In 1874 Reed & Welt planned and constructed the *Rosie Welt*, 1435 tons; she was launched at 9:45, Saturday, November 2. After spending two days on the flats, she was hauled across to Storer's wharf. The tug *Resolute* arrived from Bath to tow her to Round Pond but found it impossible to turn her in the narrow channel. The *Rosie Welt* became the first Waldoboro vessel to be towed stern foremost down the bay. Having launched their ship, Reed & Welt immediately turned to new business. The *News* reported that "before sunset the launching ways had been cleared away, and the crew had partly stretched the keel of a new 3 masted schooner, the *Eva C. Yates*. No grass grows in that yard!"

South of Reed & Welt was the yard of Alfred and son Levitt Storer. From 1830 to 1890 these men and other members of their family built some 30 vessels including the country's first 5 masted schooner in 1889, the *Gov. Ames*. In 1874 Alfred and Levitt laid the keel for a 3 masted schooner to be named *Teresa A. Keene* in October and launched her in July of the next year.

On November 6, 1874 another positive editorial appeared in the *News*: "Our shipbuilders do not appear to be depressed by any financial scare or by any cry of dull business. Three new keels have been laid within days, and there is prospect of more being stretched before spring. We hear of at least one ship as likely to be built next season together with several smaller vessels. If our capitalists pursue such a course as this, the laboring class in Waldoboro must be benefited. The splendid class of vessels which have been sent forth from our yards of late has attracted the attention of ship owners and commanders from abroad and will have a tendency to increase our shipbuilding business in the future."

How would anyone guess after reading this optimism that the industry, which had given so many hundreds of men so much for so long would be dying within a very few years?

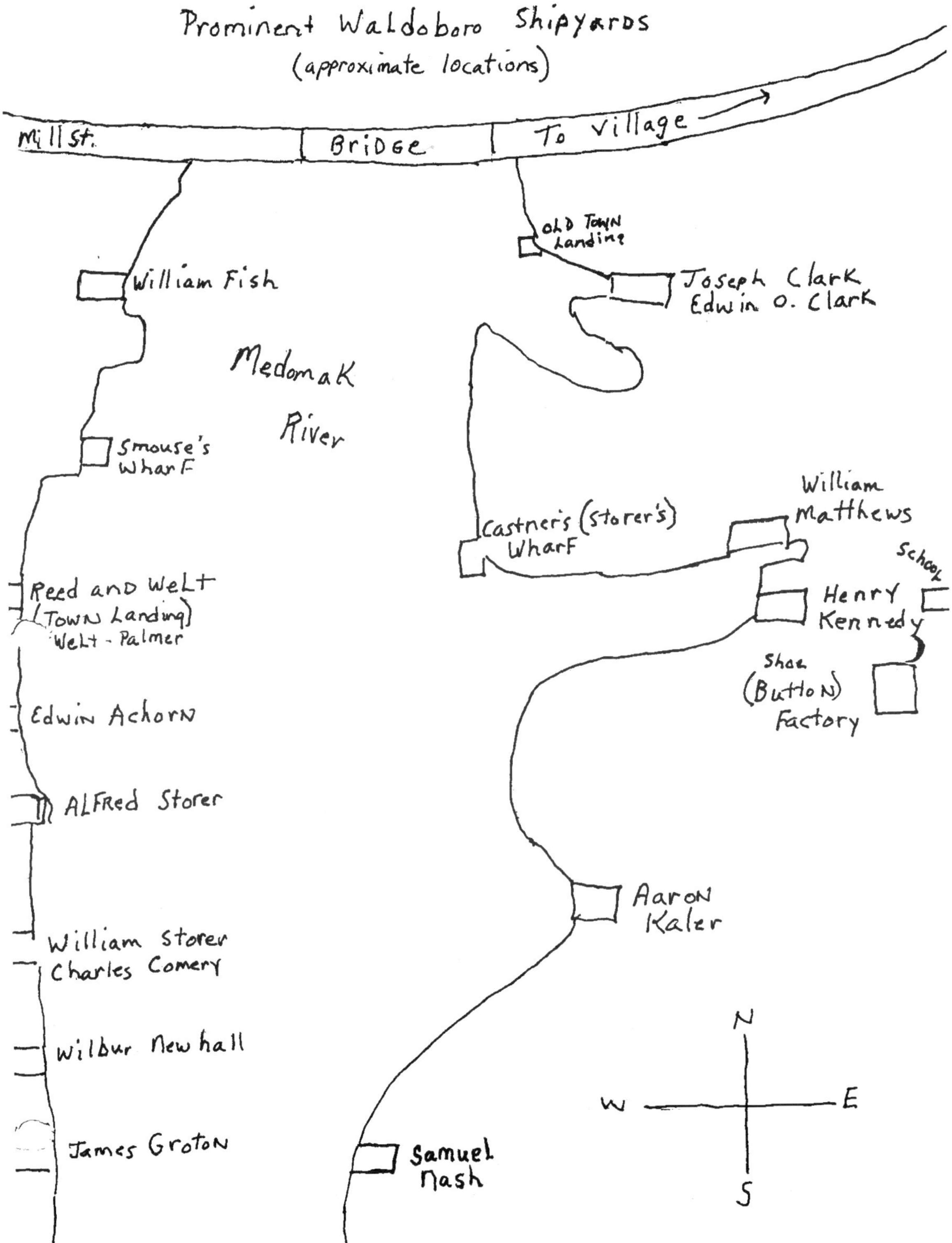

Prominent Waldoboro Shipyards
(approximate locations)
Mill St.
Bridge
To Village
Old Town Landing
William Fish
Joseph Clark
Edwin O. Clark
Medomak River
Smouse's Wharf
Castner's (Storer's) Wharf
William Matthews
School
Reed and Welt
(Town Landing)
Welt - Palmer
Henry Kennedy
Shoe (Button) Factory
Edwin Achorn
Alfred Storer
Aaron Kaler
William Storer
Charles Comery
Wilbur Newhall
James Groton
Samuel Nash
N
W
E
S

The Storers – Five Generations in Waldoboro

The banks of the upper Medomak River are very quiet today. The busy shipyards of the 19th century are gone, with many of the shipbuilding families moving away and taking their fortunes with them when hard times came. The pounding and screaming of axes, saws and hammers, the sights of veritable small armies of workers arriving for and leaving work each day have been forgotten. Brick buildings constructed during the 1850's on Friendship Street remain, as do many of the old homes. But the excitement and the drama of what Jasper Stahl calls "the Great Days" are no more.

One element of the town which connects the present with those great days directly is the Storer Lumber Yard on Friendship Street run by Alfred ("Fred") Storer, the 5th generation of his family to reside in Waldoboro. Fred's ancestor Alfred (1815–1882) arrived here, a member of the third generation from a family of German immigrants. (Johann Georg Storer, his wife Christina, and their four children had arrived in Broad Bay October 20, 1752 on the ship *St. Andrew* from Oschelbronn, Germany.) Growing up in the 1820's, Alfred, along with his brother William F. (1806 – 1881) quickly rose to become a prominent Waldoboro businessman and community leader. Together with the various partners – including Alfred's son Levitt (1850–1912) – the two men built a total of 33 vessels between 1839 and 1890. Stories of some of these ships – including the most famous of all Waldoboro craft – the *Gov. Ames*, the first five masted schooner ever, built by Levitt in 1888 are written here.

The Storer brothers' efforts earned fortunes which allowed them to invest in land on both sides of the river as well as out of town. Many hundreds of acres were purchased for woodlots in Waldoboro and a few miles north in Washington. (The Washington property by itself measured 1000 acres.) Fred tells of his ancestor Levitt (1850–1912) recounting a journey on a laboring ox-powered wagon up Main Street, listening to the church bells tolling 2 A.M., proceeding north to Washington by way of ice covered Medomak Pond, loading wagon with lumber and returning to the shipyard by nightfall. Jasper Stahl reports that in 1850 Alfred was considering purchasing a 1000 acre woodlot in either North Carolina or Virginia for "around $3000.00." Along the eastern ridge of Friendship Street the brothers built two lovely homes. William sold his in 1876 to a Dr. Eveleth; Alfred's family (Fred today) has continued to occupy his homestead.

Although the Storers' interests lay primarily in their businesses, both brothers (and their descendants) became involved with town and state affairs. Alfred, a quiet but influential man, was a lifelong Democrat (the political party of most Waldoboro voters during this period). He served two terms as Selectman – 1857 and 1858, Director of the Medomak Bank when it became National in 1865, and on many town committees. William (a Republican) served one term as Selectman – 1855 – and ran twice unsuccessfully for the state legislature – 1856 and 1863.

As previously stated, however, the Storers were businessmen first and foremost. They invested their hard earned money into their own vessels and into those of other builders – in and out of town. When shipbuilding began to falter, they moved into other local avenues of possible revenue. After the Civil War, William became a pioneer in the ice producing business. He created a pond above Sproul's Mill north of the lower bridge, had the ice cut out and stored in a sawdust insulated house on his property just south of the shipyard area on the west side. Alfred's son Levitt in 1888 bought into the Castner-Caldwell Lumber/Coal yard on the east side, eventually owning the business outright. In the latter years of the 19th century and early 20th century, it was a common sight on the river to see schooners loaded with coal unburdening at the Storer Wharf. Stan Hanna, long time resident, remembers being hired to shovel this cargo – fuel for homes and businesses of the area – from ship to shore. Levitt's male descendants (interestingly named Alfred, Levitt, and now Alfred (Fred) have maintained the business ever since. A true connection with the Great Days!

Sources for this article include: Conversations with Fred Storer, a conversation with Stan Hanna, *The Lincoln County News*, Stahl's *History of Old Broad Bay and Waldoboro – Volumes I and II*, and Whitaker and Horlacher's *Broad Bay Pioneers*.

Storer genealogy:

Original settler – Johann Georg Storer, 32, a carpenter and tailor, from Oschelbronn, Germany, his wife, Christiana (Seiter), 34, and their children Johann, 10, Matthias, 8, Ephrosyna, 4, and Anna Margetha, 2, arrived in Broad Bay about October 20, 1752, along with 127 others.

Alfred (1815 – 1882) – wives Susan (Kaler) (died 1844) and Elizabeth (Kaler) (died 1887)

Levitt (1850 – 1912) – wife Josephine (Smith) (1867 – 1943)

Alfred (1887 – 1973) – wife Francis (1895 – 1988)

Levitt (1927 – 1988) – wife Mary Ann (Parsons) (1926 –)

Alfred – Current owner of lumber yard – wife Mary (Kennedy)

Levitt – Student in 2004

William F. (1806 – 1881) – wife Susan Kensel (1807 – 1886)

William A. (1840 – 1898)

Alfred Storer 1815 – 1882

J. MANCHESTER HAYNES
Alfred Storer, 1885, 756 Tons
Courtesy Maine Maritime Museum

"The schooner J. Manchester Haynes *was built by Levitt Storer in 1885. Her dimensions were 756 tons; 179.6 X 35.6 X 20.2 feet. She was owned by the builder and others in 1885; J. Baker and Co., Boston 1892-1896; Levitt Storer and others 1900-1906; Charles F. Webb 1913. Masters of this vessel were Jonathan Mathews 1885-1896; A Marshall 1900-1906. The* J. Manchester Haynes *was abandoned at sea February 26, 1919 – Latitude 35.33 N., Longitude 70.49 W.*

Robert Applebee papers, Penobscot Marine Museum, Searsport

FRANK HARRINGTON
Alfred Storer, 1880, 634.38 Tons
160 x 33.5 x 15.6
Courtesy Maine Maritime Museum

From historian Robert Applebee's records: The Schooner Frank Harrington *was commanded by C.S. Kent 1880–1886. She was owned by C.S. Kent, J. Baden and others 1880–1886. And she was reported lost in 1890.*

Storer Family Vessels

YEAR	RIG	TONNAGE	NAME	BUILDER
1839	Ship	543	*St. Andrew*	William and Alfred Storer
1843	Ship		*Herman Hunry*	"
1844	Ship		*Carolus*	"
1851	Ship	880	*Levitt Storer*	Alfred Storer
1854	Ship	1200	*Alfred Storer*	A. Storer and J. Hovey
1855	Ship	1148	*James Hovey*	"
1856	Ship	1700	*William F. Storer*	W. Storer and C. Comery
1856	Ship	1628	*Hamilton Fish*	"
1857	Brig	300	*Ambrose Light*	Alfred Storer
1862	Brig	396	*John H. Kennedy*	Storer and Caldwell
1863	Schr	350	*S.H. Jackson*	William Storer
1864	Schr	254	*Whitney Long*	"
1865	Schr	264	*J.B. Marshall*	Alfred Storer
1865	Schr	310	*Hattie B.*	William Storer
1866	Schr	264	*Daniel Pierson*	Alfred Storer
1867	Schr	266	*Maggie Mulvey*	"
1867	Ship	1070	*Alfred Storer*	"
1867	Schr	111	*Nellie Belle*	"
1868	Schr	237	*Oriola*	"
1871	Schr	339	*Edward Emerson*	"
1873	Schr	476	*Lula*	"
1873	Schr	498	*John Proctor*	Alfred Storer & Son
1874	Schr	377	*George E. Young*	"
1875	Schr	315	*Teresa A. Keene*	William Storer
1877	Bktn	604	*Ralph M. Haywood*	Alfred Storer & Son
1879	Schr	586	*Isaac T. Campbell*	Alfred Storer & Son
1880	Schr	634	*Frank Harrington*	Alfred Storer & Son
1881	Schr	744	*Woodward Abrahams*	Alfred Storer & Son
1882	Schr	443	*George H. Ames*	Alfred Storer & Son
1883	Schr	550	*Joshua Baker*	Levitt Storer
1885	Schr	756	*J. Manchester Haynes*	Levitt Storer
1888	Schr	1689	*Gov. Ames*	Levitt Storer
1890	Schr	1064	*James W. Fitch*	Levitt Storer

William Storer – 1806–1881 – brother of Alfred; uncle of Levitt
Susan (Kensel) 1807–1886

Alfred Storer – 1815–1882 – father of Levitt; brother of William
Susan Ann (Kaler) died 1844
Elizabeth (Kaler) died 1877 – four years sick Both daughters of Capt. George Kaler

Levitt Storer – 1850–1912 – son of Alfred; nephew of William
Josephine (Smith) – 1867–1943

Ship Hamilton Fish – *1628 tons – William Storer and Charles Comery 1856*
Courtesy Maine Maritime Museum

"It is true in the greatest days of sail the largest and fastest ships were built in New York and Massachusetts. All Black Ballers were built in New York with two exceptions, and those exceptions do honor to the reputation for quality of the Waldoboro yards, for the only two Black Ballers ever built outside of New York State were the two sister ships, William F. Storer *and* Hamilton Fish. *Both were built by William F. Storer and Charles Comery in the Bill Storer yard on Waldoboro's west side. These ships were three deckers with white oak frames and of 1628 tons. They had a length of 200 feet, a beam of 42 feet and a depth of 28. Built in 1856, the* Storer *ran the Black Ball packet schedule until 1875 and the* Hamilton Fish *until 1878."*

Stahl – Volume II – pages 266 – 267

The World's First Ocean Liners

The Black Ball Line was one of several highly competitive shipping lines organized after the War of 1812. Black Ballers sailed on a set day each month New York for Liverpool; the fastest ships available were sought for this run. Previously, two months were needed for such a trip, but in the first six years of operation, the new line's vessels' average time was 23 days going east and 40 west.

Another account – that of Robert Applebee – has the following: "the William F. Storer, *one of the large transatlantic packets in the Black Ball line, was owned by Trask and Dearborn in 1861, Charles S. Marshall & Co. 1864–65, T. Dunham's Nephew & Co. 1884–93, and L. Luchenbach 1901. She was renamed* Hamilton Fish *in 1901. Her rig was changed to schooner/barge and she foundered off Barnagut, N.J. March 6, 1906."*

(Confusing!)

The Lincoln County News – May 3, 1877 – "Launched by Alfred Storer a staunch and beautiful barkentine of about 650 tons. She is named *Ralph M. Hayward* and is to be commanded by Captain I.B. Doane of Harwick, Mass.; she is owned by the captain and parties in Providence, R.I."

L.C.N., May 24, '77 – "The barkentine *Ralph M. Hayward* makes a fine appearance as she lies at Smouse's Wharf. The symmetry of the hull and the beauty of the rig is apparent."

L.C.N., June 7, '77 – While being towed to sea, the *Ralph M. Hayward* got ashore on one of the Seal Rocks in the Medomak River. Damage appears to be small."

Barkentine Ralph M. Hayward – *604 tons – Alfred Storer & Son 1877*
Courtesy Waldoborough Historical Society

From the records of historian and Customs House employee Robert Applebee:

"The barkentine *Ralph M. Hayward* (152' X 33' X 15') was owned by Joseph Baker Jr. and others in 1877, M.L. Gilbert of New London in 1910, and J. Lewis & Sons of Boston in 1918. She burned August 4, 1918 at Savanna la Mar, Jamaica."

Schooner *Isaac T. Campbell*

This photograph shows the three masted schooner *Isaac T. Campbell* at the Alfred Storer yard in Waldoboro a short time prior to her launching September 26, 1879. (The unfinished ship on the ways in the foreground is A.R. Reed's Down Easter *Emily Reed*.)

The Lincoln County News of 1879 recorded her building progress: May – keel laid; September – planking complete; September 26 – launching. After being launched, she ran across the channel and grounded up on the east side of the Medomak River. That night she slid 40 – 50 feet and ended up with her bow on the west side! She was towed to the Kennebec River on October 16.

Her dimensions were 157 feet in length, 33.9 feet in width, and 15.3 feet in depth.

Historian Robert Applebee tells us her later history: She was owned by James Baker and others from 1879 to 1893 and later by J.H. Conant and others 1893 to 1906. On August 27, 1909 she foundered 450 miles off the Louisiana coast. Two men were lost.

Isaac T. Campbell *1879*
Alfred Storer Yard, 586 tons
Courtesy Maine Maritime Museum

Stories of other vessels built by the Storer family:

Schooner John Proctor – *Three masted; 499 tons; 140' X 32' X 15.6'; built in 1873. In her 36 year career she was managed by C. and J.F. Baker. The end came when she was stranded off Cape Henlopin, Delaware, September 13, 1909.*

Ship Alfred Storer – *1116 tons; built by Hovey and Storer in 1854; commanded by Captain Charles Comery of Waldoboro for a time (see accompanying story); may have been of clipper design; in 1862 she was still half owned by Hovey interests; the other half was owned by Jarvis of Castine and Hall of Boston. Sold foreign and hailed from South Shields, England in 1866.*

Ship James Hovey – *1088 tons; built by Hovey and Storer in 1855; may also have been of clipper style. In 1864 she appears in American Lloyd's as* Margaret Quale *hailing from Liverpool. In 1886 American Lloyd's has her as* Pickforton Castle *hailing from Liverpool.*

Schooner Woodward Abrahams – *744 tons; built by Alfred Storer & Son in 1881. Owned by Harrington, King, and others 1881 – 1900, and by Fields and Pendleton of New York in 1910. She was torpedoed and lost April 22, 1917, Latitude 52 North and Longitude 29 West.*

From a speech made by Customs Agent and maritime historian Robert Applebee to the Waldoboro Lions Club May 12, 1955.

Courtesy Maine Maritime Museum

"The Big Schooner"

The town of Waldoboro produced more than 600 wooden sailing ships. Of that total, the vast majority were TWO and THREE MASTED schooners – the pickup trucks of their day. These vessels were typically from 120 to 150 feet long, 20 to 30 feet wide, and 12 to 14 feet in depth and were capable of carrying 400 to 700 tons of cargo. They were usually coastal travelers bearing anything that would pay – ice, wood and coal being most common.

Waldoboro's shipbuilding story, however, has always been highlighted by accounts of her BIG vessels, the FIVE MASTED schooners. There were 7 of these: Levitt Storer's *Gov. Ames*, and the 6 William Palmer vessels built by George L. Welt. When we say BIG, we now think in terms of 230 – 300 feet long, 40 – 50 feet wide, 22 – 27 feet deep, and with cargo carrying capacities of 1700 – 2800 tons. They were, in fact, by far the largest ships ever to be launched into the waters of the Medomak River.

It was early in 1888 that Levitt Storer and a crew of 150 began work on the first five masted schooner ever built in the United States. (Two other five masters had been constructed – one on the West Coast and the other in the Great Lakes, but neither was technically a schooner.) This vessel was named for Oliver Ames, governor of Massachusetts 1886–1888; she was to be owned by her Captain C.A. Davis and others in Fall River, and later on parties from Providence and New York.

The cost of the *Gov. Ames* was $75,000 (most ever paid for any Waldoboro craft until World War II.) She was built from 460 cubic tons of selected Virginia white oak and 876,000 feet of Georgia pine and measured 232 feet at the keel. Her lower masts were 115 feet long and 30 inches in diameter; the topmasts towered 56 feet higher, and her jibboom was 75 feet long. Under full sail she could spread 7000 yards of canvas; carrying capacity was 1764 tons.

On December 1, 1888 the "big schooner" was launched to the delight of hundreds of spectators. One early portent of bad luck occurred when John Weston, the launchmaster, lost his foot as a turn of rope twisted around it, and before anything could be done, pulled if off – so cleanly that no blood flowed from the wound. This ill luck would continue on the maiden voyage – under charter to load coal at Baltimore for Providence. Having sailed in ballast from Round Pond harbor December 9, the *Ames* was caught in a tremendous gale 50 miles southeast of Cape Cod. Captain Davis attempted to take in all sail and anchor, after bringing the ship up into the wind. It was too late; one by one the giant masts came crashing down. To make matters worse, the three ton anchor and 180 feet of chain could not hold, and the ship began to drag in the direction of George's Bank. Fortunately, two fishermen appeared and were able to remove the captain and transport him into Gloucester from which he could contact an ocean going tug. Towed into Boston the *Ames* was repaired and ready for sea in about a year – at the cost of $10,000. It was a bad beginning. One newspaper wag explained all the trouble with the comment "God didn't like to see any vessel at sea with more than four masts!" And, as it happened, no more five masters were built ANYWHERE for the next 10 years.

Despite this inauspicious beginning, the *Gov. Ames* enjoyed a very long and profitable career, all things considered. We have reports about her carrying cargoes both on the East and West coasts of the U.S.

as well as to ports in Latin America. One such story had her delivering coal to the lumber mills of Bangor and Orono in May of 1900. While tied at the pier in Bangor, the force of an ebb tide snapped her 10 inch hawser (with a report like a gun shot), and she dragged her anchor over to the Brewer shore. Next high tide she was hauled off the bank and back to a new berth where she lay "with enough lines on her to hold a battleship!"

Finally, in 1909, 21 years after her launching, the *Gov. Ames* was on a voyage from Brunswick, Georgia with 33,000 railroad ties. An enormous winter gale struck, and on December 13 at 10 A.M. the big vessel piled into Wimble Shoal, 25 miles north of that infamous oceanic graveyard, Cape Hatteras, North Carolina. Only one man, Josiah Spearing survived, clinging to a hastily constructed raft until he was picked up by a steamer bound for Charleston, South Carolina. He was eventually able to tell the story of the *Gov. Ames*' final resting place.

(Author's note – Mr. John Gatcombe of Waldoboro has constructed a lovely weathervane/model of the *Gov. Ames* under sail. It is displayed at the site of the original building ways – the Town Landing.)

Sources used in the writing of this article were Stahl's *History of Waldoboro*, a speech by historian Robert Applebee, an article by historian Mark Hennessey in the *Portland Press Herald*, and *The Lincoln County News*.

THE GOVERNOR AMES.

The Great Five-Masted Schooner Successfully Launched.

Saturday was a gala day for Waldoboro. Hundreds of people came from far and near to witness the launching of the leviathan five-masted schooner Governor Ames, which has been building since the latter part of last July. The lines for this huge piece of marine architecture were taken from a model by Albert N. Winslow, designer, of Taunton, Mass. The dimensions of the Governor Ames are as follows: Length of keel, 232 feet; length over all, 265 feet; breadth of beam, 49.6 feet; depth, 21.2 feet; lower hold, 13 feet; between decks, 8 feet. Her gross tonnage register will be 1778.77; net 1689.84, and her coal carrying capacity 3000 tons on a draft of 20 feet, light water draft, 9 feet.

In the construction of the Ames 460 cubic tons of Virginia white oak timber of the best quality went into the frame. The floor timbers are sided 13x14, and moulded at 17 inches at the breast or keel. In ceiling, planking and decks 800,000 feet of the best quality of Georgia pine were used. The main keelson is 8 feet deep, constructed of 14x14 timber. The sister keelsons are the same as main keelsons. The floor ceiling is 12 inches in thickness, the bilge keelsons 14 inches, the lower deck clamps 12 inches, the lower deck beams 11x14, the upper deck beams 12x14.

The outside planking of the Ames is 6 inches thick and at the gunwale it is 7 1-2 inches, which is the heaviest planking ever put on to a schooner. The centreboard is 33 feet long, 15 feet deep and 9 inches thick, and it is made of the best quality of white oak.

In the fastening of the schooner, which was under the direction of John W. Creamer, a master fastener of 30 years' experience, 150 tons of bolt iron were used. In the main and sister keelsons 40 tons of 1 3-8 and 1 1-4 bolt iron were driven, and in the floor ceiling 30 tons of 1 1-4 and 1 1-8 bolt iron were used from keelson to lower deck clamps. The between decks fastening is 1 1-4 and 1 1-8-inch bolting. The ceiling is edge bolted from the main keelson to the upper deck clamps. From the outside the schooner is butt bolted from floor timber heads to gunwale through and through, and in addition 10 streaks on bilge are butt bolted every sixth timber through and through from luff of bow to bends aft. In this fastening 1-inch galvanized iron was used. The master fastener says that in all his work as a master fastener he never before saw so much iron put into a vessel of her tonnage.

For the loading and unloading of cargo the Ames has nine hatchways, each 24 feet across decks and eight feet fore and aft. The cabins are large, the main cabin being finished in oak and antique woods and the other in natural woods. The schooner's quarter deck will run forward about 60 feet, some 15 feet forward of the main cabin.

The five lower masts—foremast, mainmast, mizzenmast, jiggermast and spankermast—are each 115 feet in length and 30 inches in diameter. The topmasts are 56 feet in length. The foretopmast is 22 inches in diameter at the cap. The jibboom is 75 feet long and 22 inches in diameter at the bowsprit cap. The schooner will spread 7000 yards of canvas made of the heaviest quality of duck.

The schooner's two anchors weigh 6500 and 5000 pounds respectively, and 80 fathoms of two and a quarter inch chain will be attached to each. She is also supplied with a stream anchor of 1350 pounds and a kedge of 1000 pounds. The Ames has a No. 11 windlass—the largest built—from the American Windlass Company of Providence, and a large steam capstan—from the same manufacturers—the latter a new departure on board schooners. A hoisting engine of 30 horse power will hoist the sails.

The Governor Ames was built to carry large cargoes, yet her lines, which are clean fore and aft, indicate that she will prove a very fast sailer. This large and only five-masted centreboard schooner will be commanded by Captain C. A. Davis, of Somerset, Mass., late in command of schooner William P. Wood, and who personally supervised the construction of the Ames. The master builder was Leavitt Storer, who has gained the reputation of building none but the best. The managing owner of the Governor Ames will be J. H. Davis of Somerset, Mass. The hailing port has not yet been decided upon by the owners. Completed and ready for sea the cost of the schooner will be $75,000.

The new schooner is known as the Governor Ames Her owners are Oliver Ames, Fred L. Ames, and Oakes A. Ames of Easton, Mass.; William T. Hart, John Goldthwaite, Willard Dalrymple and Henry C. Delano of Boston; S. H. Tingley, D. W. Davis, George Bucklin, John Heathcote of Providence; George H. Chase, Edward K. Morse, Captain E. G. Davis and J. B. Westgate of Fall River; A. Longstreet of Chicago, Albert H. Winslow of Taunton, Hon. F. H. Stevens of Swansea, N. R. Davis & Son of Assonet, and William P. Hood, J. F. Davis and C. A. Davis of Somerset.

Previous to eight o'clock a large crew had been industriously driving the wedges and removing the obstructions, and at eight o'clock the great structure slid gracefully into the Medomak, where she was taken in charge by tug Amelia, of Bath, and towed down the river about four miles where her sails will be bent and a few finishing touches added. The launching was witnessed by hundreds of people, about seventy-five going from this city on the early morning freight. There were on board the Governor Ames as she slipped from her cradle about ten gentlemen and two ladies, a FREE PRESS reporter being among the number, besides the workmen. He also enjoyed the delightful sail down the river and a pleasant chat with Captain Davis, who has the distinction of being the commander of the only five masted schooner ever built. He is a middle aged gentleman, a skilled navigator and a very successful master mariner.

The accommodations of the Governor Ames are all that could be desired. Her spacious cabins are elegantly fitted and arranged very conveniently. They are finished in polished oak, and will be beautifully furnished.

The Governor Ames will sail for Baltimore in a few days.

Clipping from the
Waldoboro Free Press,
December 5, 1888
Courtesy of Fred Storer

RESCUED AT GEORGES.

HOW TUG MORSE BROUGHT BACK THE GOVERNOR AMES.

An Advertiser Reporter's Trip On the Stout Towboat—The Danger in Which the Schooner Was Placed—The Method of Rescue and the Return.

The largest tug in the world towing the largest schooner in the world—a five-master, dismasted on her first voyage—was what the watchmen at Chatham saw early Thursday morning, and in a few minutes reported in Boston, tug H. F. Morse, Capt. Blair, of Bath, having in tow the five-masted schooner Governor Ames, Capt. C. A. Davis, of Somerset, which was dismasted off Cape Cod and drifted to the Georges, 120 miles from land.

At 10:15 A.M. yesterday the heavily laden steamship Chatham, Capt. Howes, of Savannah, bound to Boston from Baltimore, passed the two laboring vessels off Nauset, near enough to Cape Cod to see the breaking spray on the beach below the sand cliffs.

Snugly ensconsed in the pilot house of the magnificent tug boat, which deserves the name of steamship if a vessel of her tonnage (198.6t gross) ever did, were Capt. C. A. Blair, Mate F. E. Lewis, both of Charlestown, Pilot F. E. Hopkins, a sturdy, red-faced down-easter of Trenton, Me., Capt. J. F. Davis of Somerset, familiarly known along the coast as Capt. "Joe" Davis, part owner of the Governor Ames and brother of her captain, who was on her when she was dismasted, and last and perhaps least, a reporter of The Advertiser, the only representative of the press on the rescuing expedition. Capt. Davis told the story of the wreck to The Advertiser reporter as together they came from George's Bank.

The Governor Ames was launched at Waldoboro, Me., December 1 of this year, fully rigged, a picture which would delight the eye of the most experienced sailor and the most inexperienced landsman alike. She set sail from Waldoboro on the 9th for Baltimore, to load with coal for Providence. On Tuesday at 10 P.M., with the wind blowing a "young hurricane" from the southeast, the splicing of the standing rigging of the foremast drew out on both sides. The fault was that the ends of the wires of which the rigging was made were cut too short in the splicing. The rigging having given way, the mast, which was 115 feet long and 29 inches thick, had nothing to support it in a gale of wind but its own strength. After swaying about for over an hour, in which time all the lower sails were taken in, the foremast went by the board, breaking off close to the deck, which left the four other masts all loose. In a very few minutes the other masts followed, pulled down by the springstays, crashing down upon the deck, and bringing with them a perfect network of wire rigging, so thick that one could hardly crawl through it.

Three of the masts fell on the deckhouse, the jiggermast breaking in two over the stern. The mizzen and spanker masts were saved, so that by splicing they can be used again. Four of the masts broke short off at the deck; the fifth, which stood in front of the donkey engine house, broke, as if by miraculous foresight, above the house, so that it did not crush the engine in falling. If it had the vessel would have been very difficult to handle, as the engine was very useful in hauling in the rigging, half of which was saved. The engine was also useful in [illegible]

On Thursday a fishing schooner came along and her captain boarded the vessel, but could give no assistance. On Saturday the fishing schooners Ethel and Maud and Governor Butler came up and got out hawsers to tow the Governor Ames. They towed her 20 or 30 miles from Saturday afternoon to Sunday afternoon, then the Governor Butler cut the hawser and sailed east. Evidently the Governor Butler was incompetent to help the Governor Ames. The Ethel and Maud, however, stood by the Governor Ames in spite of the Butler's trick of cutting loose from both the vessels. The Governor Ames then cast anchor on the stormy Georges.

As the little fishing schooner could do nothing alone, Capt. J. F. Davis, with two ship joiners who had gone with the ship to finish the work on the cabin, got on board her and sailed to Gloucester, reaching there Monday at 10 A.M., after a 14-hours' run, the fastest on record. He paid $600 for the passage, as she had not done any fishing and her trip was worth about that amount. He took the first train for Boston, reaching there at 3 P.M., and went immediately to the custom house, where Collector Saltonstall and his assistants did everything in their power to get communication with the revenue cutter Albert Gallatin. Capt. Davis and the collector telegraphed to five or six harbors along the coast, receiving return telegrams from all but one or two places.

On Tuesday, the next day, at 1 P.M., Collector Saltonstall received a letter from Capt. Abbey of the cutter, stating that he was away when the telegram was received, but it would not have made any difference as it was storming badly and he would not take out the cutter in a storm. Besides, he said, the schooner had undoubtedly got out jury masts and was working its way in toward the coast. After the cutter had got down its light spars and had coaled up, he said, he would go out and look for the schooner.

After Capt. Davis saw the letter he determined that his dependence must be on tugboats, not on cutters. If the cutter captain expected to meet half way an immense schooner working in toward the coast under jury masts in the teeth of a northwest gale he had not much dependence on the captain's judgment.

He immediately went around to the different tugboat offices and engaged the H. F. Morse, a tugboat that was not afraid of anything, to go after the schooner.

Meanwhile, the 18 men, to say nothing of the vessel, were being pounded about on one of the most dangerous places on the coast, where, as a fishing schooner captain expressed it, he would not stay at anchor for $500 a day, while the letter of the cutter captain was slowly coming from Wood's Holl, that snug harbor, where light spars can be easily taken down.

As soon as the H. F. Morse could be coaled and made ready for sea she sailed for the Georges, having on board as passengers Capt. Davis and The Advertiser reporter, who was present through the kindness of Capt. J. T. Morse, the Boston agent of the Morse Transportation Co. The tug set sail from Gove's coal wharf, East Boston, about 8:30 P.M. under a cloudless sky, the full moon and the stars making the harbor very light. It sailed past the frowning forts, past vessels lying at anchor, past buoys that told of dangers to be avoided. Right in midchannel, abreast Fort Warren, was seen a light on the top rigging of a fishing schooner run down while lying at anchor by the steamer Glaucus. The places of other wrecks, inside the harbor, were also pointed out. Boston light was passed at 10 P.M. and Cape Cod light at 1 A.M., an average of 10½ knots an hour. Long before this, however, the reporter was abed and sleeping, with somewhat troubled dreams. "At midnight there was a great cry heard;" the waves were rolling rather high, rocking the tug through angles of 70 to 90 degrees, and one wave, rather more frolicsome than the rest, spilled out Pilot Hopkins, a sailor and the writer upon the floor. After enduring the swell for a few minutes, the only landsman aboard paid tribute to Neptune in the time-honored way. Capt. Blair encouraged said landsman with stories of the sickness of a reporter of the New York Herald, who went with him in the B. W. Morse after the Leary raft, and who was sick all the time, "bleeding over a quart from his lungs;" moreover, predictions as to the way the waves would run on the Georges, and doubts as to the reporter's existence [illegible] time, with a northwest gale liable to drive them into the broad ocean, to die of starvation and thirst! It meant help, home, friends, plenty!

At noon we reached her and could see the men working on her deck. The waves were rolling her as she lay at anchor head toward us, at times the whole deck almost being in sight and at other times only the bow being visible. A large flock of gulls were eating slush and other refuse thrown from the schooner in her lee. We went round under her stern and soon gave them to understand that she should send us a line. A small rope was fastened to two billets of wood and let out from her stern; this was caught and a hawser fastened to it. The schooner's donkey-engine quickly brought our hawser on board and half an hour was spent in getting out its kinks. Then the schooner's anchor was brought to the surface and the long homeward trip was begun. At 3:30 P.M. another hawser was fastened to the first, making about a 250-fathom line, because the weather looked somewhat threatening, and two vessels only 960 feet apart are a little too close in case of a storm. Nothing of moment occurred until Thursday at 7:30 A.M., when the land off Chatham was sighted, and we were reported in Boston. Before this a four-mast schooner with all sail set, a three master and a double sticker were seen. At 10:15 A.M. the Chatham passed us.

There was only a swell now to remind us of old ocean's power. At 12:30 P.M. we passed Cape Cod light, two miles away. The life-saving station man ran up flags which stood for the letters R. T. M., which mean in the international code "Glad to see you!" We answered with the saluting whistle. At 3 P.M., off Race Point, the ocean being very calm, the tug stopped, the hawser was shortened, the anchor of the schooner, which had been dangling below the water for a day, was hove up to the cathead and Capt. Davis went aboard the schooner. While we were engaged in this work along came the Gallatin, trolling its log behind, and spoke the schooner and tug. What the people on the schooner said is not recorded here and let us hope an angel did not record it above. The Gallatin told us that the tug Argus had gone out from the Vineyard on the same errand as ourselves yesterday noon. Then the cutter crossed our bow and headed for Boston, followed by the mocking refrain, "If you get there before I do, just tell them that I'm coming too."

At 3:45 we sighted the steamer Shawmut, having in tow barges Robena and Joel Cook. At 4 the sun set over an ocean so still that you could see it slowly sinking like a ball of gold in a basin of quicksilver. Above, the sky was almost cloudless. Indeed, with the exception of the wind and swell, the whole trip was as pleasant as could be desired, not a sign of rain, and only a very light squall of snow falling during the 40 hours trip. Almost all the time an overcoat was a useless encumbrance except when on the lookout.

At 4:30 P.M., while the officers' mess were eating and swapping long yarns, Minot's Ledge light was sighted, and at about 5 o'clock Boston light and Long Island light also came in sight.

At 6:30 P.M. Minot's Light was abeam, and soon the lights of vessels became so frequent that they looked like a chain of many colored beads, with here and there a break in it along the horizon. One set of beads seemed to come nearer and nearer, and finally Capt. Blair said, "It's a tug that wants to speak with us." Nearer and nearer it came without a word being said, till an active man could have jumped from our deck on to the pilot-house of the stranger. Then the whistling and loud talking that followed would have astonished a Quaker or a woman. The tug fell back and spoke with the schooner about the question of anchoring, being answered that the Morse Co. had charge of the whole business. The tug was not satisfied but came back to us again, straining and puffing to catch up, and was again answered in the same way.

The Glaucus, outward bound, gave us a salute, which we returned. We were now nearing Boston Light, when another tug came up. This was the Samuel Little, belonging to the Morse company. She stuck by us all the way through the Narrows. Near Fort Independence a coasting schooner came near us, and by the rule of the road at sea we had to let her pass, between Nix's Mate and the gas buoy, before [illegible]

WALDOBORO'S BIG SCHOONER

Damaged by the Gale and Anchored off the "Georges."

A Nor' Easter Would Drive Her on the Shoals.

GLOUCESTER, Dec. 17.

J. F. Davis a passenger on the schooner, Gov. Ames, Waldoboro for Baltimore, on his arrival here today on the schooner Ethel Maud said the Ames called at Round pond, sailing from there Sunday. Had moderate weather till Tuesday, when she encountered a gale 50 miles southeast of Cape Cod carrying away the foremast. The rigging became loose and all the other masts went by the board some falling aft on the deck. The weather was very cold and the vessel became unmanageable.

She drifted for some hours when she was fallen in with by the schooners Ethel Maud and Gov. Bat er, who were unable to tow her in. Soundings were made on the Georges, the anchors let go, and the vessel brought up in twenty fathoms of water.

When Davis telegraphed to Collector Saltonstall to send a cutter to the Ames' assistance, the vessel was in a good condition, having plenty of provisions with Captain Davis and fourteen men on board. They were cheerful and expected to get her into port. Captain Buckhart of the Ethel Maud, said he entertained no fears for her safety so long as the wind kept to the northward and westward, but she anchored in a bad place being close to Georges Shoals. Should the gale come from the northeast it would be impossible for fishing vessels to render any assistance towards towing her, she being of such enormous size. It is understood that the Ethel Maud get $600 for coming in for assistance. Considerable rigging and sails were saved. Davis said there was only a small insurance on the vessel. He left for Boston this afternoon to have an interview with the collector to go out in a cutter.

Life of a Shipyard Worker

The fortunes to be made and the lovely homes built in Waldoboro were for the very few. While the "Dons" of the village – Welts, Reeds, Storers, Kennedys, Clarks, and others lived in luxury because of ships and shipbuilding, most of the hundreds of working men lived a much different life. (Needless to say, very few of their homes exist today.)

The six day week of the shipyard worker began very early. From dawn to dark was the set schedule until about 1850 when ten hour days were instituted by unions. (One owner was heard to say that he didn't mind ten hour days, as long they were ten hours before lunch and 10 hours after lunch! And many employers flouted "the rules".) In any case, if one had to travel to the job as many as 10 to 12 miles, one had to begin early. After 1875 trains were available, and many used them. But for most men travel was by horse, buggy, boat or foot. Some of the yard owners saw to it that housing was available in the neighborhood for some of the longer distance travelers.

As most of the work was done out of doors, the weather had to have been a factor. But rarely were businesses closed because of heat, cold, rain or snow. On July 4, 1878, *The Lincoln County News* reported that the men of Henry Kennedy's yard had been sent home July 2 when the temperature reached 99 degrees in the shade! This was very unusual; the *News* pointed out that a yard in Damariscotta had taken the same action, and that after all, this was the worst heat wave in 30 years! When the work was available, the yards were open – all year around.

Once on the job, the men were divided into groups or crews under a foreman and under the general supervision of the Master Builder or Superintendent. Sometimes these crews worked permanently at one yard; sometimes they were subcontractors who moved around from ship to ship. For example, we find the Davids Brothers rigging crew employed at many places – yards in Waldoboro but also in Damariscotta – Newcastle over the years. Each crew did over and over the specialty they had been trained to do: planking, caulking, painting, rigging, sailmaking, shipsmithing, joinering, fastening, etc. An individual was responsible for the care of his own tools and for his midday lunch – a gargantuan affair of a half dozen sandwiches, pie, coffee etc. eaten somewhere on the grounds in the hour afforded. There was no safety equipment and no insurance for injury on the job; there was no protective rain gear provided – in fact, rheumatism early in life was often a result of wearing wet clothing for hours at a time.

When the day was over, the men would return home often to do several hours of chores on the farm before retiring to bed. Children were ordered to be especially quiet at night so that their fathers could sleep. (Fathers and their children obviously saw little of each other except on Sunday.) Early bedtimes also meant a precious saving of fuel in the cold seasons. For the most part the shipyard worker lived in a fairly primitive fashion near the poverty line. Diseases such as measles, cholera, smallpox, and consumption (tuberculosis) were all too common especially with the young. And with the pay averaging out to about $2.00 – $2.50 a day (by 1889)*, doctors and other expensive luxuries were difficult to afford by many.

Having painted this rather grim picture, it is well to remember that being employed at a shipyard – particularly at one of the more skilled jobs – was most desirable. The money was the best that could be earned in the area at the time, and the conditions were the same everywhere. Men were not apt to complain (at least too loudly) about their lot.

The following are some glimpses into life in the Waldoboro shipyards as reported by *The Lincoln County News*:

May 1, 1874 – "Rufus Achorn lost part of his thumb in the circular saw."

May 8, 1874 – "Snow 3 feet deep in Waldoboro."

"Ira Welt had a narrow escape Thursday. The discharging schooner *Wild Rover* lost a block strap, and he was struck on the head. Prostrated by the blow and thought to be badly injured, he was found by Dr. Eveleth to have a one inch scalp wound and contusions on his arm."

May 29, 1874 – "A bell has been placed upon the blacksmith shop in the Clark yard by William Waltz. This will announce to all yards the hours for commencing and knocking off work. It will be rung at 6:30, and 11:30 A.M., and at 12:30 and 5:30 P.M.

August 21, 1874 – "During the launching of *Josephine*, Joseph Wall (colored) was thrown into the river by one of the stern lines, and in great danger from drowning and from flying timbers. Joe came here at the closing of the War (Civil) and by good conduct has become a favorite with his employers and acquaintances."

October 16, 1874 – "More than 100 men are employed at the Clark shipyard."

October 23, 1874 – "William H. Mueller fell into a small reservoir for water at Reed and Welt's steam box at an early hour yesterday breaking one of the bones of his left arm about half way from elbow to wrist."

January 4, 1877 – "Labor in the shipyards has been principally limited to snow shoveling for the past two weeks."

January 25, 1877 – "Edwin O. Clark has a shipyard crew busy cutting pine lumber in Warren."

May 15, 1877 – "Sampson's sparmakers are hard at work at the Railroad Depot for A.R. Reed where the spars were delivered Wednesday."

October 11, 1877 – "The ship *Mabel Clark* has been furnished with a steering apparatus from the James P. Boyd Iron Foundry, Jefferson Street, Waldoboro."

* In the *Report of the Industrial and Labor Statistics* for shipbuilding in the Waldoboro district, the following facts were available for neighboring towns Camden and Thomaston in 1889. Note that pay had improved a good deal from earlier days when a dollar for dawn – to dark labor was the rule.

Average daily wages: Foreman – $4.00; Ship's Carpenter – $2.00 – $2.25; Joiners – $2.50; Blacksmiths – $2.50; Fasteners – $2.00; Riggers – $2.00; Sparmakers – $2.50; Sailmakers – $2.50; Painters – $2.00; Laborers – $2.00.

One Autumn – Four Schooners

Of the 500+ wooden sailing vessels built at Waldoboro the three masted schooner was the most popular. Schooners were the practical, economical "pickup trucks" of the sea, contracting to carry any cargo – from lumber to ice and food – and carry it anywhere – from coastal U.S. ports to other coastal ports (very common) or even across oceans at times. Schooners required very small crews (therefore very small payrolls), and, because of the way they were rigged could sail close to the wind ("tack") – an advantage in coastal travel. They were known as "gaff rigged" sailers as opposed to "square rigged" sailers (such as brigs and ships, who were better able to sail "off" or "down" wind but were not able to tack well and also needed much larger crews).

The autumn of 1882 saw four three masted schooners launched in Waldoboro. At this time five village yards were busy – on the east side just below the bridge E.O. Clark was employing 65 men. A bit further down river, just beyond the present day Storer Lumber Yard, the Henry Kennedy family employed 28 men. On the west side, opposite Clark, was the William Fish yard where 35 men worked; at the present Town Landing was the Augustus Welt – A.R. Reed establishment, and then – a stone's throw south – lay the yard of Alfred Storer & Son which employed 31. Business was so good that every one of these places was planning the construction of two ships – not just one – in rapid as possible succession.

As we detail the launching statistics, the reader should notice the fact that several subcontractors – riggers, sailmakers, sparmakers, etc. were often found working on more than one vessel at a time. In fact the David Brothers rigging crew did ALL of the work for ALL four schooners and "not a word of dissatisfaction was heard!"

Alfred Storer & Son launched their schooner first – on September 1. The *George H. Ames* was 140' X 34' X 11'; her net tonnage was 443. The master workman was John Storer; heads of work crews included Joiners (finish carpenters) – S.O. Waltz & Sons, Sparmakers – Wilfred and Judson Mank; Caulkers – Randall Oliver; Riggers – David Brothers; Painters – A.A. Newbert; Blacksmiths – J.S. Hatch. The ship was owned by the builders, the Storer family; James A. Bliss & Co.; J.P. Elliot of Boston; Waldo Stewart of Providence; and Dr. George H. Ames of Providence. Her captain was Alvano Marshall of St. George. Launching was overseen by the very experienced John H. Weston. Two weeks later the *George H. Ames* was towed to the Kennebec River where she was to take on a cargo of ice.

The next vessel launched was on September 29 at E.O. Clark's. Witnessed by several hundred people the *Nantasket* slipped into the Medomak River flawlessly. She measured 145'X3513' and 522 tons. Superintendent (Master workman) – Horace Flanders; Head Joiner –William Waltz; Spars – James Sampson; Painting – Osgood Miller; Rigging – David Brothers; Sails – Lane & Jones. *Nantasket's* frame was Virginia white oak; Hyde of Bath furnished windlass and pumps; she was steered by James Boyd's best machinery. Her cabin was furnished in solid black walnut and ash with gilt ornaments, elegant Brussels carpets adorned the floor, and the sofa and chairs were covered with velvet. No wonder when she appeared at the Haywood and Gage Ice Wharf in Savannah, Georgia a few weeks later she was described by the local paper as "one of the most handsome vessels we have seen." She was owned by the builder, Mr. Clark, and also by John S. Emery of Boston, parties in Providence and Portsmouth, and by her captain, Edward A. Richardson.

The Henry Kennedy yard next launched their schooner *Annie B. Hoffses* on November 3. The *Hoffses* measured 428 tons and was 141'X34'X12'. Her Master Builder was William A. Young; Head Joiner George Storer; Blacksmiths Waltz & Son; Ceiling (inside planking) was done by Charles Waltz; Spars by James Sampson; Riggers were David Brothers; Sails by Lane and Jones; Power Capstan and Pumps were provided by Knowlton; and James Boyd supplied the steering mechanism. The *Hoffses* was owned by the builder, the Kennedys, as well as by parties in Boston, New York, and Philadelphia and by her captain, David Hoffses of South Waldoboro. The vessel was designed for general freighting; her first charter was to proceed to Baltimore and load coal for Galveston, Texas. She sailed out of the Medomak River, never to return, on November 15.

Finally, on December 1, into an icy river Mr. William Fish launched the last of this foursome, the "staunch and handsome" *Stephen G. Loud*, 474 tons – 144'X34'X13'. In this venture the Master Workman was Lewis K. Benner; Head Joiner Lewis Keizer; Blacksmith George D. Sides; Sparmaker James Sampson; Caulker Miles W. Standish; Painter F.O. Miller; Riggers David Brothers; Sailmakers Lane and Jones. The *Loud* was owned by the builder, Mr. Fish, parties in St. George, Boston, and New York and by her captain, Isaac Fountain of St. George. her charter was to take a cargo of lumber from New York to Trinidad; she arrived in New York December 8.
(The *Stephen G. Loud* was still plying the waters of the Waldoboro area in 1894, at that time owned in Thomaston.)

A busy and productive shipbuilding season had come to an end in Waldoboro. But at the Clark, Kennedy, Storer, and Fish yards the unfinished hulls of four more schooners lay ready for work in 1883. Times were good.

Information for this article was obtained from the files of *The Lincoln County News*, J.J. Stahl's and SF Miller's *Histories of Waldoboro*, and a customs house document found at the Waldoboro Public Library.

Joseph Clark – Shipbuilder (Part 1)

On July 23, 1875, *The Lincoln County News* published the obituary of Joseph Clark – "an extremely serious calamity for the business interests of Waldoboro." Mr. Clark was well known not only as a very wealthy man but also as a man of the highest business principles. The largest tax payer by far in town at the time of his death, he left a fortune of about three quarters of a million dollars (worth many times that amount today).

Born in Jefferson in 1799, he was raised on a farm by his uncle after both his parents had died. As a young man he had taught school for a short time in Jefferson (and enjoyed in later years pointing out to friends and family the schoolhouse at which he labored). In 1823 he removed to Waldoboro and in 1824 took on two partnerships: one – Mary Ann King of Whitefield for his bride and the other his cousin Henry Kennedy on whose father's farm he had grown up, as a business associate. Three years later he and Kennedy went their separate ways with no hard feelings both to enter the burgeoning ship building trade. Mr. Clark's marriage prospered, Mrs. Clark having time to act as the cook in the basement of their first home on Pleasant Street for his hungry crew of shipwrights, as well as the mother to five girls and one boy. (The boy, Edwin O. joined his father at the shipyard beginning in 1872 and continued the business successfully on his own after his father's death until 1883. The daughters – Carrie, Celeste, Ella, Mary, and Caroline were all active in the theater (Clark's Hall) that their father built for them – and the town – on Main Street). Mother Mary Ann died in 1854.

Joseph Clark's first vessel, the topsail schooner *Aramanth*, was built at a leased yard just south of the present Storer wharf in 1830; his second across the river at the present Town Landing site. He then was able to purchase the land on the south side of Main Street including waterfront just below the bridge on the east side. From this site he launched some 40 vessels over the next 40 years. Struggling at first – a singularly self made man – he persevered with indomitable energy and tenacity – often daring to try what other men might not. Soon his shipyard was prospering, giving work to hundreds over the years. When the terrible fire of 1854 destroyed much of the town, Mr. Clark was among those who unselfishly helped pay for the rebuilding some of the brick blocks on Friendship Road that are still used today, as well as the Customs House (Library).

Two anecdotes give us some understanding of Mr. Clark's nature. The first glimpse comes from the story of the ship *Caroline and Mary Clark*, built in 1849. After some incredibly bad luck in the building and launching (which would have discouraged many), the vessel was towed down river. There is a hilltop south on Friendship Road known in those days as "Clark's Lookout". On this spot Clark and other shipbuilders would station watchmen who could view the lower river bay and waters beyond. When a vessel was spotted they would ride pell mell back to the village and report. What Joseph Clark's lookout saw and reported one morning was that the brand new ship *Caroline and Mary Clark* had become hung up on the eastern side of the famous "Narrows". Soon a great hole was worked into her bottom, and the possibility of a catastrophic loss was imminent. The news spread throughout the village, and sympathetic citizens made their way to the Clark office with long faces and dire questions and answers about what to do. Finally, Mr. Clark had had enough of their gloom and doom. "Let her go; let her go!!" he roared. "I'll build another ship – just like her!" (As it turned out, insurance more than covered the needed repairs, the ship was saved and paid for herself in the very first voyage!) Good fortune had smiled, as she very often did, on Joseph Clark, but typically, he had prepared himself not to look back and wring his hands over failure but to forge ahead.

Another anecdote illustrates more of the character of Mr. Clark. In the days just prior to the Civil War, he took a most unpopular stand in the eyes of Southern cotton merchants and Maine ship owners and builders. Declaring that he supported the Union and its new president, Abraham Lincoln, he infuriated many of his associates whose vessels transported this cotton throughout the world; they called him a "Black Republican". He and his friends retaliated against their Southern sympathizers by calling them "Copperheads". One day a huge American flag was hung across Main Street – from Clark's Hall on the east to the Fish Block on the west. When some local Copperheads talked about removing it, Mr. Clark took action. Arming himself with a rifle, he stood on a platform in front of his building. There he remained all day after announcing with conviction that if anyone molested the flag, "he would get a bullet in his hide." The flag flew until sundown – the hour appointed to lower it. People knew that Joseph Clark was a man of his word.

In a forthcoming article we will discuss Mr. Clark again – this time concentrating upon some of the vessels he built.

Information for this article comes from *The History of Old Broad Bay and Waldoboro* by J.J. Stahl, from *Waldoboro, Maine* by S.L. Miller, and from *The Lincoln County News*.

Joseph Clark 1799 – 1875

Joseph Clark – Shipbuilder (Part 2)

Between 1830 and 1874 Mr. Joseph Clark built more than 40 sailing vessels – all but two of them in the little cove just south and east of the Main Street bridge. A complete list of them is available (along with those of his son, Edwin O.). However, our space is limited, and this account will concentrate on seven of these. Suffice to say, Clark ships were to be found sailing all over the globe from Waldoboro to the South Seas.

Innovation and daring were two of Mr. Clark's characteristics. His third project, the *Mary Ann*, was the first full rigged ship to be built in Waldoboro. Previously, only small sloops, two masted schooners, brigantines, and brigs had been constructed – 1783 – 1835. Larger vessels which could perhaps venture further were considered gambles. *Mary Ann* of 1835 was a gamble and a successful one. Thirty years later Joseph Clark gambled once again. Up to this point, all schooners built in Waldoboro were two masters. In 1865 he ushered in the post Civil War period by constructing the *American Eagle*, a three master, and the first of its style to be built locally. Unfortunately, she was a very unlucky ship. On launching day she refused to move down the ways and was not in the water until nightfall under Captain Michael Singer. Soon after she went to sea, another skipper, Eugene Wade of Waldoboro, relinquished his command when he observed rats leaving the ship in Boston. And on her maiden voyage – Harve, France to New Orleans the *American Eagle* was tragically lost with all hands. At a tiny place called Cat Island the dismasted vessel appeared weeks later, only to be drawn back to sea by a hurricane. The next day much of her cargo – 50 French immigrants now corpses – drifted ashore and had to be buried. The hull of the *Eagle* eventually floated up on the island, and the timbers were used by the natives for housing.

The term "Down Easter" refers to large, sturdy full rigged ships built after the Civil War usually for transoceanic trade. Two of these were built by Joseph Clark. The *Gold Hunter*, 1258 tons, was built in 1867 to the order of J. Henry Sears of Boston and proved herself to be a profitable ship during her 13 year career. She made passages in the Cape Horn trade, the transatlantic trade and carried coal or case oil to China. In spite of two disasters – colliding with the ship *Anahure* in San Francisco Bay and then having her rudderhead twisted off when rounding Cape Horn and having to continue her voyage to San Fancisco under tackles, she remained active. In 1878 she was in the news when her crew mutinied in Shanghai. She was finally lost when, laden with coal and bound from Cardiff, Wales to Hong Kong, she struck a reef in the South China Sea.

In 1874, one year before his death, Mr. Clark teamed up with son Edwin to build the firm's largest vessel. The *Carrie Clark*, 1326 tons, was owned in part by her builder and also by Boston parties for about 9 years. In 1883 this Down Easter was sold to the Germans and renamed *Anna*. After operating for about 20 years under the German flag, the ship was purchased by Lewis Luchenbach of New York, converted into a barge, and her original name restored. *Carrie Clark's* career came to an end after 47 years of sea life (of which 29 had been under canvas). While carrying coal along the Atlantic coast, she foundered off Highland Light, New Jersey on November 28, 1921. Generally considered "a good and reliable vessel and a good carrier, if rather slow," she had proved to be well modeled and never experienced any major mishaps.

The Lincoln County News – November 12, 1880 – E.O. Clark received a telegram telling of the total loss of the ship Gold Hunter. *The crew was saved; 5/8 of the ship was owned in Waldoboro."*

Several Waldoboro ships carried eager gold seekers to California before and after the great discovery in 1849. Two of these vessels were built at the Clark yard. The *Caroline Clark*, 800 tons, built in 1836 (another source says 1844) sailed with 38 passengers under the command of Captain Bradley and made a passage to the gold fields in 190 days – November 16, 1849 to June 14, 1850. The bark *Louisiana*, 299 tons, built in 1845, accomplished this feat in a shorter 161 days with 40 passengers and 12 crew, December 23, 1848 – June 2, 1849.

In a previous story we told about the ship *Caroline and Mary Clark* (800 tons, 1849) and how, after her launch she became hung up on the Narrows Rocks down river. BEFORE her launch, Mr. Clark and his men had to endure an almost unbelievable series of setbacks. 1) When the crew at the yard attempted to raise the stern into place, the block and tackle broke. The stern fell and cracked. 2) While raising the bowsprit, the strapping broke. 3) While raising one of the masts, the shear leg support broke. 4) When launching on Thursday, the ways spread, making it necessary to launch on Friday, traditionally an unlucky day. Finally she was towed down river – and went ashore on The Narrows, punching a hole in her and causing Joseph Clark to utter his famous order: "Let her go; let her go – I'll build another – just like her!" But, as previously recounted, the *Caroline and Mary Clark* was saved, insurance salvaged the damages, and she paid for herself in the very first voyage! She had a successful career, eventually being "sold foreign", renamed *Elena*, and hailing the Mediterranean island of Majorca as her home after 1866.

Self-made, daring, innovative, independent, loyal to town and to his family, Joseph Clark is an unforgettable name in the history of Waldoboro.

Information for this article was taken from: Samuel Miller's *History of Waldoboro*; Jasper Stahl's *History of Old Broad Bay and Waldoboro*; Fairburn's *Merchant Sail*; Goodman's *Encyclopedia of the California Gold Rush Fleet*; and *The Lincoln County News*.

The Lincoln County News – November 24, 1882 – "An account from the new Tacoma, Washington Weekly Ledger carried the following story: 'At Port Townsend (Washington) 13 sailors from the ship *Carrie Clark* were arrested and charged with mutiny for refusing to sweep down the deck and square the yards. Their reasoning was that they had been up all night, and it was Sunday. Captain Storer was forced to hire substitutes to do the work of getting the vessel under way. Upon arrival at Port Hudson, he had the sailors tried; the verdict was in Storer's favor – a compromise was arranged with the 13 sailors paying the costs ($160.00) of hiring extra men. W.F. Jones represented the defendants; Mr. Osborne the plaintiffs.'"

Ship Carrie Clark – *1298 tons – Joseph Clark & Son – 1874*

The Lincoln County News – January 24, 1878 – "We are pleased to see Captain Ira Storer, home for a short time. Captain Storer is one of our own Waldoboro boys, who worked his way up to his present position by perseverance and ability. His ship, the *Carrie Clark*, is currently in New York for a China port – probably Hong Kong."

Barkentine Josephine – *598 tons – Joseph Clark & Son – 1874*

The Lincoln County News – January 24, 1878 – "Mrs. Stahl, wife of Captain J.B. Stahl of the barkentine *Josephine*, in jumping from the vessel to the wharf in New York broke her leg just above the knee. The accident will detain the lady for about 6 weeks."

The Lincoln County News – March 1, 1877 – "The barkentine *Josephine*, Captain John B. Stahl, arrived safely at Otaga, New Zealand, December 9."

Clark Family Vessels

YEAR	RIG	TONNAGE	NAME	BUILDER
1830	Schr	116	*Amaranth*	Joseph Clark
1833	Brig	175	*Grand Turk*	"
1835	Ship		*Mary Ann* *	"
1836	Ship	800	*Caroline Clark*	"
1837	Ship	480	*Avon*	"
1839	Ship		*Mallabar*	"
1840	Ship	598	*Sea Lion*	"
1841	Ship	625	*Desdemona*	"
1845	Bark	299	*Louisiana*	"
1846	Schr	132	*Bay State*	"
1847	Ship	723	*George Evans*	"
1849	Ship	800	*Caroline & Mary Clark*	"
1850	Brig	175	*Edwin*	"
1851	Ship	925	*R.B. Sumner*	"
1853	Ship	998	*Ella A. Clark*	"
1854	Brig	200	*Shibboleth*	"
1856	Ship	1308	*Joseph Clark*	"
1859	Ship	1318	*Loch Na Gar*	"
1859	Ship	1308	*J. Webster Clark*	"
1862	Ship	1240	*Otis Norcross* +	"
1863	Ship	782	*Edwin Clark*	"
1864	Bark	737	*Celeste Clark*	"
1865	Schr	386	*American Eagle*	"
1866	Bark	516	*William Brown*	"
1867	Ship	1258	*Gold Hunter*	"
1868	Brig	233	*Isaac Howland*	"
1870	Brig	336	*Joseph Clark*	"
1872	Bark	782	*Nina Sheldon*	Joseph Clark & Son
1873	Bktn	484	*Mignon*	"
1874	Bktn	598	*Josephine*	"
1874	Ship	1326	*Carrie Clark*	"
1877	Ship	1661	*Mabel Clark*	Edwin Clark
1879	Brig	373	*Stacy Clark*	"
1882	Shcr	496	*Nantasket*	"
1883	Schr	404	*Evie B. Hall*	"
1883	Bktn	655	*E.O. Clark*	"

* First full rigged ship built in Waldoboro
+ Later name changed to *Gebroader Van Walig*

Joseph Clark – 1799 – 1875 – father of Edwin, Caroline, Mary, Celeste, Carrie, Ella
Mary Ann (King) – 1803 – 1854

Edwin O. Clark – 1839 – 1917 – son of Joseph; father of Stacy and Mabel
Mary (D.) – 1830 – 1897

More Clark Genealogy:

Caroline Clark – 1825 – 1888 – daughter of Joseph
Alden Jackson – 1810 – 1877

Mary E. Clark – 1851 – 1923 – daughter of Joseph
William Allen – 1855 – 1907

Celeste A. Clark – 1837 – 1929 – daughter of Joseph

Carrie Clark – 1839 – 1895 – daughter of Joseph

Ella A. Clark – 1832 – 1904 – daughter of Joseph
Albion Oakes – 1827 – 1859
Carrie Oakes – 1856 – 1886 – daughter of Albion and Ella – granddaughter of Joseph
C.R. Manville – husband of Carrie
Minnie Oakes – died June 7, 1858 age 2 months – daughter of Albion and Ella

Joseph W. Clark – 1827 – 1863 – son of Joseph and Mary
Abby – 1827 – 1855

Col. Atherton Clark – 1826 – 1882 – son of Joseph; grandfather of Maude Clark Gay
Mary D. – 1830 – 1897

Stacy Clark – son of Edwin; grandson of Joseph

Mabel Clark – 1873 – 1930 – daughter of Edwin; granddaughter of Joseph
Dr. Warren Sanborn – 1863 – 1934

Maude Clark Gay – 1876 – 1952 – granddaughter of Col. Atherton; great granddaughter of Joseph
John Gay Jr. – 1873 – 1937

Ann Gracia Gay Bailey – 1900 – 1975 – daughter of Maude C. Gay and John Gay Jr.
Stanley Bailey – 1902 – 2001

The Vessels of Samuel Nash

As indicated in the introduction, there are many Waldoboro ships and shipbuilders about whom we have very little information. One of these is Samuel Nash who may have journeyed from Nobleboro to the eastern shore of Medomak to carry out his business. We do have a list of the vessels he constructed, and some genealogy – both kindly contributed by a descendant, Mrs. Delma Baldwin of Providence, Rhode Island – and we have two stories compiled by Robert Applebee, historian and ex-customs house agent. The list and stories follow:

YEAR	**RIG**	**TONNAGE**	**NAME**	**BUILDER**
1832	Schr	117	*Packet*	Samuel Nash
1833	"	112	*Jane*	"
1833	"	147	*Vanda*	"
1834	"		*Sarah Nash*	Saml. Nash; B & J Eugley
1836	"	118	*James*	Samuel Nash
1839	"	130	*Edward Kent*	"
1839	"	130	*Van Buren*	"
1841	Brig	250	*Ohio*	"
1844	Schr	169	*Sarah Nash*	"
1845	"	127	*Algonia*	"
1846	Brig	259	*Susannah*	"
1847	Schr	149	*Albano*	"
1848	Schr	162	*Jenny Lind*	"
1848	Brig	134	*Samuel Nash*	"
1849	Schr	77	*Denmark*	S. Nash & R. Miller
1849	Schr	144	*Robert Miller*	Nash & Miller
1850	Bark			Samuel Nash & Co.
1852	Brig	232	*Peerless*	Samuel Nash
1854	"	258	*Henry*	"
1855	"	270	*Ocean Wave*	"
1855	"		*Francis Geyer*	"

Nash Genealogy
Church Nash – 1745 – 1794 – from Scituate – Samuel's grandfather
Eve Christian Reed

Samuel Nash Sr. – 1781 – 1867 – Samuel's father
Margaret Sprague

Samuel Nash – the builder – 1805 – 1892
Sally Sprague

The story of the brig *Henry*, built in 1854 – 104 feet long, 26.7 feet wide, 10.6 feet deep. She was owned by Samuel Nash 3/16; Isaac Umberhind 1/16; John W. Tibbets 1/16; James Ludwig 1/16; Nathan Nash 1/16; Robert Nash 1/16; William S. Cochrane 1/16 – all of Waldoboro; Roland Jacobs of Thomaston 1/16; Hiram Robinson 1/16; Edward Gilchrist 3/32; Peter Morgan 1/32 – all of St. George; Ambrose Snow 1/32; Joseph Burgess 1/32; A.W. Badger 1/32 – all of New York.

(Ownership of vessels in this manner was typical in shipbuilding towns like Waldoboro. Sometimes, the vessel was very lucky and paid for herself after the first voyage. Sometimes, and tragically, she was lost on the first voyage, ruining some of her hopeful investors.)

The last trip for *Henry* was in 1857 when she was three years old and sailing from Aspinwall to St. Domingo. When she arrived on December 11, Captain Charles Farren was very sick with fever and subsequently died at sea. Mate Peter Morgan, next in command, then came down with the same malady and died December 26; he managed to appoint Seaman John Clark acting master before expiring. Despite heavy weather Clark worked the ship into Cartagena, New Grenada, December 30. But the storm had crippled the *Henry*, carrying away her sails and masts, and the crew found it impossible to make the harbor entrance. They tried an alternative route – Boca Grande – but found it temporarily closed because of low water. (*Henry* drew eight feet; there were only three feet of depth when they approached.) But somehow, on the crest of a gigantic wave, the little brig made it to safety. Once in the harbor, Clark dispatched four of the crew to New York on the brig *Townsend Jones* while the remaining two stayed on board to care for the ship. Shortly thereafter the *Henry* was sold to foreigners at the port of Cartagena.

Another Samuel Nash vessel, the 86 foot X 23.7 X 8.4 foot schooner *Albano*, built in 1847, was commanded in 1847 by Hiram Pitcher. She was owned by her captain and, like the *Henry*, many others – including nine from Waldoboro (Alfred Storer and George Kaler amongst them), three from Friendship and one from Thomaston.

Henry Kennedy 1799 – 1875

House of Henry Kennedy – Main Street – Waldoboro Bicentennial Book 1973

Deacon / General Henry Kennedy

In an earlier Waldoboro Wanderings account we described the death and enormous funeral service of Joseph Clark, shipbuilder and benefactor extraordinary for Waldoboro. Just three months after the town paid its tributes to Clark, another giant of the town, Henry Kennedy, also passed away. For in October of 1875, Kennedy, while superintending some repairs on one of his properties on Friendship Road (then called Water Street) fell through the rotten boards of the roof. 15 feet below he landed, breaking his collar bone and several ribs and sustaining many bruises. He lingered a few days in great pain before finally slipping away October 13. On the following Sunday all services at all the other churches were suspended as most citizens of the town (and many from away) crowded into the Baptist sanctuary on Main Street. The Reverend L.D. Hill, assisted by the Reverends McLeod and Simonton, performed the memorial service.

Henry Kennedy had been born in the northwest part of Waldoboro on a farm September 23, 1799. He worked for his father, attended some school and then eventually made the decision in 1824 to leave the farm for the exciting young village and join his cousin, Joseph Clark, another farm boy from Jefferson as a trading partner. After two years both young men yearned to be on their own; the shipbuilding business beckoned the ambitious all up and down the Maine coast. Eventually each had his shipyard – both on the east side of the Medomak River with Clark's place just north and Kennedy's just south of the present day Storer Lumber Company.

Starting in 1837 Henry Kennedy built over 30 sailing vessels – ranging in size from the little (103 ton) *Henry A.* (still to be seen in the village harbor 100 or more years after her launching in 1851) to the 820 ton full rigged ship *Moonlight*, his largest. After Henry died in 1875, sons Henry and L.L. (Lincoln) continued the business under the firm name H. Kennedy & Co. Henry the younger and L.L. were two of the five children father Henry and wife Rachel (Lincoln) raised in the big white house across the street from the current U.S. Post Office.

As mentioned, the Kennedys attended the Baptist Church (Henry having converted from the more established Congregational Church.) In 1844 he gained the title "Deacon" because of his enthusiastic and active participation. He was equally helpful in town and state affairs and held many offices: hog reeve, clerk, selectman, and customs collector. After losing as an unpopular Republican candidate for State Senate to Isaac Reed in 1862 – 501 votes to 2 (!), he won a term in the Senate after the Civil War brought the Republicans into prominence.

"Deacon" Kennedy was also involved in a near war with Great Britain. In 1839 both U.S. troops and British troops were rushed to the Maine – New Brunswick border. The treaty of 1783 which supposedly set the border lines proved ambiguous, and Great Britain claimed territory all the way to the Penobscot River, while the U.S. claimed land all the way to the St. Lawrence River! Men rushed to "join up", and one of the hundreds was Deacon Kennedy. Happily, bloodshed was averted when the Governor of New Brunswick promised not to occupy the land in question. American troops were withdrawn – including Kennedy who had risen to the rank of Brigadier General. Forever afterwards he would be referred to as both "Deacon" and "General". But the so called "Bloodless Aroostook War", fortunately was no more than that – bloodless.

Shipbuilder, churchman, public servant and generous employer – the loss of Henry Kennedy on the heels of the death of his first cousin Joseph Clark was a body blow to Waldoboro.

Sources for the information in this article came from Stahl's and Miller's *Histories of Waldoboro* and Zetz and Zoidis book *Woodsmen and Whigs, a History of Bangor*.

Ship Moonlight *– 820 tons – Henry Kennedy – 1855 Courtesy Peabody Essex Museum*

From Robert Applebee's records: The Moonlight *was reported abandoned December 1877.*

Kennedy Family Vessels

YEAR	RIG	TONNAGE	NAME	BUILDER
1837	Schr		*Columbia*	Henry Kennedy
1838	Schr		*Moscow*	"
1841	Brig	279	*Toronto*	Kennedy & Welt
1846	Schr		*Mary H. Cheppell*	Henry Kennedy
1848	Bark	385	*Amazon*	"
1848	Schr	141	*Alabama*	"
1849	Schr	139	*Martha Hall*	Kennedy & Hall
1850	Schr	56	*Canary*	Henry Kennedy
1850	Ship	449	*Oregon*	"
1851	Schr	103	*Henry A.*	"
1851	Brig	199	*Three Sisters*	J.R. Kennedy
1851	Brig	563	*New England*	Young, Kennedy & Co.
1851	Brig	195	*Lucy Ann*	Henry Kennedy
1851	Ship	744	*Toulon*	"
1855	Ship	820	*Moonlight*	"
1855	Brig	231	*Almore*	"
1857	Brig	346	*Fannie Lincoln*	Kennedy & Hall
1863	Brig	480	*Sarah F. Kennedy*	"
1864	Brig	392	*Edwin H. Kennedy*	"
1866	Bark	750	*Laura A. Kennedy*	"
1867	Bark	248	*Elodia A. Kennedy*	"
1868	Schr	234	*Zeta Psi*	"
1873	Schr	449	*Ella M. Storer*	Kennedy & Comery
1874	Schr	435	*Mary J. Cook*	H. Kennedy & Co.
1875	Bktn	448	*Florence L. Genovar*	"
1878	Bark	758	*Fannie L. Kennedy*	"
1880	Bark	711	*Matanzas*	"
1882	Schr	404	*John H. Cross*	"
1882	Schr	428	*Annie B. Hoffses*	"
1883	Schr	407	*Ida Francis*	"
1883	Bktn	639	*Rachel Emery*	"

Deacon (also General) Henry Kennedy – 1799 – 1875
Rachel (Lincoln) – 1812 – 1892

L.L. (Lincoln) Kennedy – 1830 – 1903 – son of Henry
Elodia (Storer) – 1823 – 1890

Semanded Kennedy – 1833 – 1834 – daughter (?) of Henry

Henry A. Kennedy – 1835 – 1884 – son of Henry

Almore – 1839 – 1913 – son of Henry (lawyer – Colby grad – lived in Main St. house)

Ghost Ship of the Medomak

I went looking for a ghost the other day; haven't found her yet but will continue the search in hope of being successful. "Her" is the schooner *Henry A.*, built in 1851 and, some say, still existing on the eastern shore of the Medomak in Waldoboro village.

The walk from the Brick School House on Friendship Street to the river is a ten minute one – the path takes the visitor through a thick growth of trees, and it winds back and forth to offset the steepness. When the river bank is visible, one can scan the eastern shore north to the Storer Lumber Yard, south downstream, and west across to the Town Landing. One hundred fifty years ago on this property the land was clear, and it was the site of the Kennedy Shipyard – Henry Kennedy first and his son "L.L." later. Thirty-one sailing vessels – schooners, ships, and barks – were built here (from 1837–1883), 60 to 100 men were employed, teams of horses and oxen worked, and sounds of pounding hammers, screaming saws, and chopping adzes filled the air. Vessels loaded with lumber and other materials arrived each week for this and the other yards. Today the bank is beautiful and silent except for the wind and the few seabirds fluttering about. One iron ring set in a stone and no doubt used for a pulley or guy wire reminds the visitors of the former great industry.

In the 1850's, when the *Henry A.* was born, Waldoboro was en economic center of the region – and the country. A national survey indicated that the town ranked seventh in the U.S. (behind New York, Boston, Philadelphia, Bath, Baltimore, and New Orleans) in total tonnage for vessels registered, enrolled, and licensed. The docks and graveled streets were crowded; population was at an all time high of over 4,000. The decade was the village's most successful one ever – in business terms. And wooden ship building was the reason.

The little schooner *Henry A.* – 78 feet long by 22 feet by seven feet deep, 103 tons burden – was apparently built to be one of the supply vehicles which brought goods called for from town to town along the east coast. She also appears to have spent her very long career coming to and from Waldoboro itself. *The Lincoln County News* reports: September 18, 1874 (the first year of this local paper) – "The *Henry A.* with her Captain Wade left town for Boston."
December 18, 1874 – "Navigation on the Medomak to the wharves was closed on the 11th; though it had frozen over once or twice previously, it had broken up, and ice carried out on the next tide. The *Henry A.* was cut through the ice down river from Light's Ferry nearly to Storer's Wharf Sunday. The extreme cold of Monday night has effectively closed the channel for the winter."
August 16, 1875 – "The schooner *Henry A.* is lying at St. George for repairs, much stronger than before her mishap." (Author's note – the "mishap" was not explained.)
July 31, 1879 – "The *Henry A.* has been repainted."
June 2, 1882 – "The *Henry A.* has discharged a load of masts for the Kennedy Yard."
December 5, 1882 – "The *Henry A.* lays at Storer's Wharf."

By 1894, at the age of 43, the little schooner was still sailing. (Most wooden ships lasted no more than 15 years.) But her active days were nearly ended.
On August 6, 1896 the *News* commented: "The schooner *Henry A.*, heretofore laying waterlogged at the Kennedy Wharf, has been raised by empty kerosene barrels and floated over the Smouse's Wharf (Author's note – present Town Landing) where she rests."

Eventually the vessel was returned to the Kennedy Wharf, and here she remained. Several present day citizens remember being told, as children, about her. Richard Wallace's grandmother, a graduate of W.H.S. 1915 when it was in the brick Friendship Street building, recalled the dilapidated old hulk sitting near the shore when she and her schoolmates were playing at recess. Fred Storer's aunt Marion remembered children swimming from the vessel in her youth.
In the 1950's, when Jasper Stahl published his history of Waldoboro, he commented about the "rotting timbers of the *Henry A.* lying off the rotting wharf."

Today the 152 year old vessel, the last of the great Waldoboro fleet, may still be somewhere on that shore. She is a ghost well worth meeting and remembering.

Sources for this story: *The Lincoln County News*, Stahl's *History of Waldoboro*, 1856 *Annual Report on Commerce and Navigation*, Maine Maritime Museum, and Waldoboro citizens Randy Gross, Stan Hanna, Fred Storer, Martha Boggs, John Murray and Richard Wallace.

Site of the Kennedy shipyard – Courtesy Waldoborough Historical Society

The Master Craftsman – Augustus Welt

When Augustus Welt, the junior partner of the famous shipbuilding team of Reed and Welt died, May 13, 1892 *The Lincoln County News* described him in flowing detail:

"He was a man of great energy, and sagacity, upright in all his dealings, a just and considerate employer, an excellent citizen and widely known for his many acts of benevolence and charity." (Author's note: Obituaries of the day tended to be a bit overblown – surely Mr. Welt had made an enemy or two somewhere!)

Be that as it may be, Augustus Welt's career accomplishments speak for themselves; he was surely one of the most respected and successful characters in Waldoboro's history. Born September 12, 1809 he grew up in Waldoboro, the second son of John W. Welt, and began his adult life rather inauspiciously: "When I was 21, my father gave me my time (freedom), a suit of clothes, and a yoke of oxen. I swapped the oxen for a horse, and the horse died. My start in life!" But life had to go on. He soon went to work as a laborer in the shipyards where the hours stretched from dawn to dark six days a week. Natural skill and energy in time earned him a promotion in the rapidly growing industry to master workman or foreman; in 1835 at the age of 24, he raised the funds to build his first vessel in partnership with his cousins, Charles and Solomon Shuman.

Five years later he and two other cousins, George Smouse and Isaac Reed the younger entered into a union which was to become one of the major town firms (along with those of Joseph Clark and Henry Kennedy on the east side of the Medomak and the Storer family on the west). This company – known henceforth as Reed and Welt was to build more than 30 ships on the site of the present Town Landing and flourish for 23 years, bringing wealth and fame to its owners. After 1876, the company disbanded, Mr. Reed out of the business and his son, A.R. continuing on his own. Augustus Welt built several more vessels – mostly medium sized schooners such as the 140 foot *Fannie Child* pictured on page 46 and his last, the four masted *Augustus Welt*, in 1889.

In November of 1877 the *News* reported that the semi-retired Mr. Welt and his daughter-in-law had left town for New Orleans where they were to meet up with the full rigged vessel *Rosie Welt* and her captain, Orlando Welt, the builder's son, for a family reunion.

Augustus Welt did not confine himself to ships and shipbuilding. (In the division of duties he had been the hands on builder, while Mr. Reed worked on the business of engaging buyers, sailing contracts, etc.). He found time also to give to his town and state. In 1838 he and 21 other men formed the first fire company and were trained to use the "Water Witch," a fire engine bought with private funds. He served two terms as first Selectman and four terms in the State Legislature. He was a Deacon in the Baptist Church, a Director of the Waldoboro National Bank and a Director of the Knox and Lincoln railroad.

Mr. Welt lived with his wife Sedonia and later on second wife Clara in their house on the corner of Medomak Terrace and Pine Street. This place was the location for three interesting incidents in town lore. In 1807 (before the house was built) a large wooden platform was erected for the purpose of ordaining John Cutting, First Congregational minister. Later, in 1842, crowds gathered here and directly across the river to witness the foretold "End of the World," when only the faithful would be raised to the heavens from earth rather than taste Death. (It didn't happen, of course, and the incident ever after was known as "The Great Disappointment".) Finally, in the 1860's, the house was a center of a spiritualism craze. The Welts and others engaged with mediums to observe table tippings and dish rattlings by supposedly cooperative spirits from the next world. One meeting found the participants of a seance rushing home in a terrible fright after a table followed the medium around the rooms and would not halt!

Sources for this article include Stahl's *History of Waldoboro*, *The Lincoln County News*, A List of Marriages from the Vital Records of Waldoboro, and Applebee's List of Vessels Built in Waldoboro.

Welt Genealogy:

Augustus Welt – 1809 – 1892 – son of John Welt (lived on Pine and Medomak Terrace)
Sedona (Cedona?) – 1817 – 1881

Capt. Orlando Welt – 1837 – Dec. 12, 1902 (lost at sea) – son of Augustus Welt
Sarah Welt – daughter of Augustus – unmarried
George L. Welt – 1855 – 1917 – nephew of Augustus Welt

Augustus Welt 1809 – 1882

Fannie L. Child *1881*
Augustus Welt & Co., 425 tons, 140 x 32.1 x 11.8
Courtesy Maine Maritime Museum

From historian Robert Applebee: The *Fannie Child* was owned by A. Welt and others 1881–1902. Her master was Isaiah Hart in 1881. She was reported lost 1902–1903.

The Lincoln County News – July 6, 1902 – "The schooner *Fannie Child* brought cargo of hard pine to the village shipyards this week."

Harold Castner, Lincoln County historian – "The Augustus Welt*, captained by George Sproul much of her career, was in the Chesapeake Bay in 1892, awaiting orders to unload her cargo. Being towed in with her was another vessel, and the officials on shore left the decision as to who would be unloaded first to Captain Sproul and the other skipper. The two commanders agreed to flip a coin; Sproul won and brought his vessel close to the dock. During the night the dock caught on fire, and before the* Welt *could be drawn away, she was in flames and burned to the water's edge. She later was towed to Boston where it took a year to rebuild her. From that time on Captain Sproul would never again toss a coin to settle any question!"*

Robert Applebee – "The four masted schooner Augustus Welt *was built in 1889 by George Welt, nephew of Augustus. She measured 203.9 X 40.9 X 21.5 feet; her tonnage was 1162. She was owned by A.R. Reed and others 1889–1901; Harrington, King and Co. 1907–1913. She was torpedoed and sunk July 21, 1917 by a German submarine after having been active during World War I, Latitude 40.30 N., 7.20 W."*

The Remarkable Reeds (Part 1)

Waldoboro's 19th century history is notable for the accomplishment of several families whose members by hard work, luck, and shrewdness gained and held power in town business, economics, and politics. In the next three articles we shall attempt to trace one of these families, the Reeds, who lived and worked here from 1807 to 1919. It is a most fascinating story – that of Isaac Reed the Elder (known as "The Colonel"), Isaac Reed the Younger or "The Honorable Isaac Reed", and the grandson Asa Redington (or A.R.) Reed.

Colonel Isaac Reed (1783–1847) was born in Littleton, Massachusetts and, after having passed the bar and become a lawyer, migrated to Maine – first to Jefferson in 1807 and then to Waldoboro in 1808. He married the widow of a wealthy merchant, George Smouse, who was also related to another well established family, the Millers. A very large, unfinished mansion on Jefferson Street was purchased and completed, and it was here that Isaac and Jane Reed were to live and bring up their large (six children and one step daughter) family.

Isaac, while establishing himself as a successful lawyer, also became very involved with military affairs and was appointed a colonel in the Waldoboro militia. In 1812 a second war with Great Britain had erupted, and feelings were running high. In 1814 a British fleet appeared off Camden; the Waldoboro troops were called to the Town House and, with Col. Isaac at their head prepared to march. Followed by weeping wives, mothers and girlfriends, they struck out, down over the Main Street bridge and up the hill heading for Warren, where they spent the first night, Rockland where they spent the second night and finally to join up with other volunteers at Camden. After a week of anticipation during which no shots were fired, they returned to Waldoboro, camped out at the Reed Mansion over night and then dispersed at Smouse's Grove across West Main Street from the Town House.

After the war Reed worked very hard at the founding of Maine as a new state. He served as a member of the Maine Constitutional Convention and conceived the idea for the new state Seal. It was in his home on Jefferson Street that Bertha Smouse, young stepdaughter of the Colonel, sketched out his design for that seal.

Colonel Reed, as he was ever afterwards to be called, found the time, also to become closely connected with town affairs. A Selectman for 40 years, a Post Master for 10, he was also a founding member, a generous contributor, and the first pew holder in the Congregational Church, newly built (near the site of the present A.D. Gray school). In politics he maneuvered himself into a position of power in the very conservative Federalist Party and saw to it that Waldoboro voted with the Federalists much of the time. When the Federalist control nationally waned, and the new Democrats under Andrew Jackson took over, Reed still maintained the conservative control in Waldoboro until, in the late 1830's and early '40's old age began to take its toll on him. Retiring to his lovely home he passed the last few years peacefully and died in 1847.

The Colonel's shoes and powerful position were not empty for long. His oldest son Isaac (later to be known as Honorable Isaac) knew from a very young age about power in Waldoboro. Born in 1809, he was educated at Bloomfield Academy but, unlike his father, with his Harvard education, preferred the opportunities in commerce to law and did not attend college. Tall, handsome, ambitious and personable, he saw the shipbuilding business as a place to make much money, and he went for it. At the same time (age 20), he took over the leadership of the conservative political machine that his father had once headed, now labeled as the Whig Party. By flattery, charm, advice, and dangling chances for employment he was able to become even a more powerful "Boss" than his late father had been.

(To be continued)

Colonel Isaac Reed 1783 – 1847

Colonel Isaac Reed's Mansion on Jefferson Street
Courtesy of a Reed descendant, Sandra Denman

Interior of the Reed Mansion – Courtesy of Sandra Denman

The Remarkable Reeds (Part 2)

As a political boss Isaac Reed the younger controlled the voting to an extent hardly to be believed today. He was known to say in private that the people were "happiest when governed by the GOOD, the WISE, and the RICH." And on another occasion in private when asked about the Waldoboro voters, remarked "I own them." The people, for their part, after being personally cultivated, befriended, advised, and often employed by Mr. Reed, waited for him to tell them how to vote. Very occasionally someone went against the boss's wishes and was punished by the loss of a contract or a job if he were employed, or by property damage if he were not. Waldoboro, as a result, voted for years extremely conservatively and often completely differently from the communities around it.

At age 26 young Isaac Reed entered into his first shipbuilding partnership – with Bela B. Haskell. After two years the firm dissolved, and Mr. Reed became senior partner in the famous and long lasting company of Reed and (Augustus) Welt. From 1843 to 1876 this team built some 40 vessels at the old Smouse – Reed place – now the Town Landing. Reed – Welt was extremely successful; their ships were sized from small coasting schooners or large full rigged ships for the most part made much money for the builders and the owners, many of whom lived far away from Waldoboro.

Young Isaac (now with his father gone known as Honorable Isaac Reed) along with his business and politics took on many town and state offices. For 32 years he served as President of the Waldoboro State and National Bank; for 25 years he was Town Moderator; for eight years a Selectman. He also worked as a state legislator for six terms; as a State Senator for five terms and was an unsuccessful candidate for Governor (as a Whig) in 1854 – 1855. He was the State Treasurer in 1856, on the State Board of Agriculture and also the State Board for the Insane.

Once established in his business he built a fine big house on Main Street (eventually torn down to make room for the Waldo Theatre). And in 1855 he gave part of his apple orchard so that the U.S. government could build a beautiful Customs House (later Post Office and currently the Public Library).

By the 1880's shipbuilding had seriously declined in Maine. Reed and Welt were dissolved in 1876; the two old partners content to slide gently into old age. Asa Redington Reed, oldest child of Isaac's, continued the industry at the same site on the west side. With his second wife (first with Caroline Redington had died after giving birth to Asa) Lydia McDonald, young Isaac, like his father, had raised another large Reed clan. He spent the last few years enjoying his back yard garden, communing with old cronies, and finishing each day with a snack of apples, popcorn, and cider. He died in 1887.

The career of Asa Redington Reed, the Colonel's grandson will complete the story of this remarkable Waldoboro family.

Main Street, Customs House, Honorable Isaac Reed's Mansion
Courtesy Waldoborough Historical Society

Honorable Isaac Reed – 1809 – 1887

The Remarkable Reeds (Part 3)

Asa Redington Reed, son of the Honorable Isaac and grandson of Colonel Isaac, was the first and only child of young Isaac and his first wife, Caroline (Redington) formerly of Augusta. Shortly after giving birth to Asa December 22, 1842, Caroline died. Isaac married again two years hence – to a Lydia McDonald who bore several more children.

Educated in Waldoboro schools and at Warren Academy, Asa very early showed none of his father's and grandfather's interest (or need) for power in politics. Except for one two year term in the Maine State legislature, he remained out of the political arena. A quiet but natural leader and much respected, he steadfastly refused also to throw his weight around in Waldoboro affairs, despite often being urged to do so.

He did follow his father into the shipbuilding business – starting his own firm A.R. Reed & Co. at the age of 25 and immediately, in 1867, constructing the *Rosetta McNeill*, a 611 ton bark. Twenty five years later he built his last vessel, *Madalene Cooney*, a large schooner. Three of the 11 vessels were named for his father, brother and sisters: the 873 ton bark *Alice Reed* in 1873, the 1550 ton ship *Isaac Reed* in 1875, the 1449 ton ship *Willie Reed* in 1877, and the 1564 ton ship *Emily Reed* in 1880. During the building of the *Emily Reed*, A.R. was employing about 100 workers in his yard. When he built the 1745 ton ship *George Curtis* in 1884, A.R. Reed could give claim to having launched the largest vessel ever (up to this time) constructed in Waldoboro. Not until the famous Palmer fleet of the early 20th century would anyone else claim this honor.

By 1890 shipbuilding was nearly at a standstill in Waldoboro. A.R. Reed, a bachelor, and retired, lived out his remaining years in the large white mansion (site of the present Waldo Theatre) that his father had built on Main Street. His connections with the Medomak National Bank – he had served as president of this organization for 23 years – continued for a while, and he enjoyed his membership in King Solomon's Masonic Lodge for sociability. Some of his friends in King Solomon's acted as escorts at his funeral after he passed away December 21, 1919.

When A.R. died, the living brothers and sisters sold their brother's house to a Robinson Cook. Later it was sold to Warren Weston Creamer, one of the first antique dealers in Waldoboro who transacted business there. Eventually in 1936 the house was sold again, this time to be torn down and replaced by the current Waldo Theatre.

Sources for the three Reed stories were Stahl's and Miller's *Histories of Waldoboro*, *The Lincoln County News*, Waldoboro Town Records, and Lincoln County Records at Wiscasset. The author also thanks Richard Wallace, Martha Boggs, Stan Hanna, Betty Lou Lee, and Darryl McKinney for their help.

Ship Emily Reed *1564 tons Built 1880 by A.R. Reed & Co. – Author's Collection*

Ship Willie Reed – *1449 tons – just launched July 1877 – A.R. Reed & Co.*
Author's Collection

Bath Commercial – July 1877 – "We occupied a short time the other day with an examination of the *Willie Reed* (in Waldoboro). We were afforded every courtesy by one of the mates. The *Reed* is a fine and substantial example of naval architecture – well and strongly built, tastefully furnished, worthy of notice in every port. Next to products of Bath shipyards, we take most pride in those of our near neighbors."

Ship Willie Reed – *Courtesy Maine Maritime Museum*

The Lincoln County News – June 14, 1877 – "A.R. Reed & Co. successfully launched the *Willie Reed* Monday forenoon. Built of white oak and hard pine, she measured 199' X 24' X 40' and was 1449 tons. She received the highest rate for her class from French Lloyd's for 11 years. Owned by Yates and Portfield of New York and others she will be commanded by Oscar Yates of Round Pond. Tuesday she was towed to Davis Point where her rig will be completed and ballast taken on."

LCN – June 21 – "The ship *Willie Reed* will transport deals (boards – usually pine or fir) from Bath to Liverpool."

LCN – July 5 – "Willie Reed, age 19, brother of the builder, intends to go to Europe on the ship named for him. A pleasant voyage to you, Will!"

LCN – August 30 – "The ship *Willie Reed* arrived in Liverpool Monday."

Fairburn – Volume V page 3381 –

"The *Willie Reed* was launched June 11, 1877, by A.R. Reed & Company and was generally considered one of the best vessels ever built at Waldoboro. She was owned mainly by her builders and Yates & Porterfield, New York, her managing owners. The ship was of 1,449 tons register, 210 ft. long, 39 1/2 ft. beam, and 24 ft. deep. Under command of Capt. Oscar S. Yates, a partner owner, the *Willie Reed* gave a good account of herself and, considering her fullness, was a good sailer. She is credited with a passage of 114 days from San Francisco to Dublin. Frederick C. Matthews writes: "On July 1, 1882, the *Jabez Howes, Seminole, Eliza McNeil,* and *Willie Reed* sailed from San Fancisco and were close together four days later. Capt. John B. Emerson of the *McNeil* wrote that his ship outsailed all except the *Willie Reed* and on another occasion when both these ships were near Cape Horn, bound to the westward, the *Reed* was given credit for sailing somewhat faster than the *McNeil*." The last completed voyage of the *Willie Reed* was a westbound Cape Horn passage of 153 days from New York to Astoria and an eastbound run of 124 days from Astoria to Queenstown. On February 24, 1894, when seventeen years old, she was wrecked on Point San Quentin near Royan, France, and became a total loss."

This painting of the Isaac Reed *was commissioned by her commander Captain Frederic D. Waldo in 1896. Waldo was the son of Captain Shubael and Martha (Haskell) Waldo and descended from General Samuel Waldo, the founder of our town.*

Information courtesy of Mr. Richard Wallace

Stories of other vessels built by the Reed family:

The bark *Alice Reed* was launched by A.R. Reed & Co. in 1873. She was 163' long and her tonnage was 873. She was named for the builder's sister Alice (1862–1914) and was owned from 1875 to 1901 by Yates and Porterfield. After a long 29 year career she was carrying a cargo of salt off Long Island, N.Y. and in a storm was wrecked December 3, 1902 on Napeague Beach. In the 1970's a young couple, Michael and Connie Hidalgo, were walking along the beach when they uncovered a metal object jutting out from the sand. Fearing that it might prove dangerous, they scraped enough dirt away to uncover the chocks of a ship which in turn were connected to 400 pounds of planking. Carved into the planking was the name *Alice Reed*. More research taught the Hidalgos the birthplace of the vessel and eventually they made a special trip to Waldoboro and to the launching site.

From an article written for the *Portland Press Herald* by Steve Cartwright May 28, 1977.

The bark *Rosetta McNeil*, 620 tons, was built in 1867 by A.R. Reed & Co., was operated by her builders, Yates and Porterfield, J. Baker, and others during her career. She was lost in a hurricane in the Gulf of Mexico in September 1882 with all hands.

Robert Applebee – Speech to the Waldoboro Lions Club 1954

Bath Times – June 1882 – "The bark *Rosetta McNeill*, built by A.R. Reed in 1867, has been overhauled at Goss, Sawyer and Packard in Bath and reclassed. Her planking and ceiling were opened, and her timbers were found perfectly sound. The vessel will probably seek a foreign service."

The *Alex McNeil* – A Down Easter – built in 1869 by Reed and Caldwell & Co. – She measured 1122 tons gross, 178 X 36 X 23 and one half feet. She was built on order for New Orleans owners and named after "Sandy" McNeil, a prominent merchant of that city. In 1888 she was purchased by J.D. Spreckels of San Francisco for the Honolulu sugar trade, and her rig was changed to that of a bark. Later she was in the export of lumber trade. On December 26, 1902, when she was 35 years old, she was returning to Puget Sound when she went ashore on Pratas Reef and became a total loss.

Fairburn Vol. V p. 3380

The *Isaac Reed* – Down Easter – was built by A.R. Reed & Co. and measured 212 X 40 X 24. She was built "on spec" and was named after the Waldoboro State Senator. She was commanded by Capt. William S. Colley, part owner, and managed by Yates and Portfield of New York. Known as an average sailer and a good carrier, the Reed was sold in 1900 to J. Jensen of San Francisco and resold in 1914 to Hind, Rolph & Co. of the same city. After being converted into a barge she foundered off the California coast in July of 1924 when she was 49 years old.

Fairburn Vol. V p. 3380
33801

ASA REDDINGTON REED

The Lincoln County News – February 27, 1879 – "Mr. A.R. Reed is taking a vacation for his health in Georgia. We hope that he finds the restorative powers that he needs."

Miss Alice McDonald Reed 1862-1914 – sister to A.R. and Willie Courtesy Waldoborough Historical Society

The Lincoln County News – March 23, 1899 – "A party of 29 took advantage of the fine sleighing and drove to Damariscotta where they dined at the Maine Hotel. Among those in the group were W.A. Richards and wife, Captain Willard Wade and wife, Clinton B. Stahl, Alice Reed, Lena Storer, and Alice Bailey."

L.C.N. – March 30, 1899 – "Miss Alice Reed entertained 13 for whist Friday night."
L.C.N. – March 30, 1899 – "At Clark's Hall Saturday night Miss Alice Reed and Mrs. Maude Gay acted out Scene I Act V from Macbeth. All proceeds were to go to help M.E. Church, Winslow's Mills."

L.C.N. – January 25, 1900 – "Miss Alice Reed has returned from a trip to Boston."

L.C.N. – February 2, 1901 – "The Moonlight Reading Club met as usual this week at Alice Reed's for the annual Valentine Party. Before lunch the group read the story 'Uncle Terry'."

From historian and Customs House Agent Robert Applebee: The schooner *Maggie S. Hart* was owned by A.R. Reed and others 1885 – 1901. Her masters were A.C. Cheney in 1885 and J.P. Farrow in 1900. She sailed from Jacksonville, Florida December 18, 1909 and was never reported.

Courtesy Captains Douglas K. and Linda J. Lee

Courtesy Maine Maritime Museum

Robert Applebee: "The schooner *Hattie P. Simpson* – 1221 tons; 210' X 41' X 21' was built in 1891 by A.R. Reed & Co. She was owned in 1913 by Alden Cheney et al. On March 21 she foundered off Cape Lookout, North Carolina, losing 6 men."

Duplicate Schooners – 1892

In June of 1892 A.R. Reed launched the first of two identical schooners for that year. *Ida C. Southard* was 178' X 37' X 18'. She was on the sea for 13 years, finally abandoned in the North Atlantic February 9, 1905 by Captain George Southard and his eight man crew. Mr. George C. Wright of Perth, Australia, sent us the report that a friend's grandfather, Mr. Stanley Cooper was in charge of a lifeboat from the British steamer *S.S. Lillie* which rescued these men. All members of the lifesaving crew as well as the skipper of the *Lillie* were given awards – a gold watch and chain for the captain, a marine glass or binoculars for Mr. Cooper, and gold medals for the seamen for their bravery – sent to them by American president, Theodore Roosevelt.

In November of 1892 Mr. Reed launched his second vessel (and, as it turned out, the very last of his illustrious career). *The Lincoln County News* reported: "A.R. Reed launched the schooner *Madeline Cooney*, a duplicate of *Ida C. Southard*. She was named for the daughter of John J. Cooney of Brooklyn, New York and measured 177' X 37.2' X 17.9'. Her tonnage is 751; George L. Welt was the general supervisor, and George Young the master mechanic. Planking was done by Wilson; fastening by Kuhn; blacksmithing by Cramer; caulking by Standish; joinering by the Waltz brothers; outboard by Benner; painting by Miller; spars by Sampson; rigging by David brothers; sails by Lane and Jones. She is owned by the builder, parties in New York and Baltimore; her skipper will be Captain Millard Wade. She is to be towed to the Kennebec River, where she will be loaded with ice."
Historian and former Customs House officer Robert Applebee says "The *Cooney* was owned by A.R. Reed and others from 1892 until 1902; C.R. Wickman owned her in 1910. On December 28, 1911 she was in a collision with *S.S. Warrington* 12 miles north of Cape Hatteras; the ship and the entire crew was lost."

Ida C. Southard *1892, A. R. Reed & Co., 818 tons, 178.7 x 37.2 x 18.3*
Courtesy Maine Maritime Museum

Schooner Madalene Cooney *– built 1892 – 751 tons – A.R. Reed & Co.*
Courtesy Maine Maritime Museum

From the records of Robert Applebee: "The *Cooney* was 177' X 37' X 18'; she was owned by A.R. Reed and others 1892 – 1902 and C.A. Wickman 1910. She collided with the *S.S. Warrington* December 28, 1911 twelve miles off Cape Hatteras. All hands were lost."

Mansion owned by Honorable Isaac Reed and A.R. Reed; torn down in 1936 and replaced by the Waldo Theatre. Courtesy Waldoborough Historical Society.

Reed – Welt Family Vessels

YEAR	RIG	TONNAGE	NAME	BUILDER
1837	Brig	189	*Virginia*	Reed, Haskell & Co.
1837	Bark			"
1843	Schr			Welt, Reed & Co.
1844	Schr	164	*Pedamonte*	Reed, Welt & Co.
1846	Schr		*Oceola*	"
1846	Schr	123	*Sarah Ann*	"
1846	Brig	169	*J.L. Whipple*	"
1846	Schr	120	*Redington*	"
1847	Bark	200	*Pedemonte*	Welt & Co.
1848	Schr	134	*Albion*	Reed, Welt & Co.
1848	Schr	149	*S.D. Hart*	Welt & Co.
1849	Brig	157	*Orlando*	"
1850	Ship	699	*Muscongus*	Reed, Welt & Co.
1851	Brig	194	*Edward*	"
1852	Ship	1248	*Edward Stanley*	Hovey, Reed & Co.
1853	Schr		*Moro*	Reed, Welt & Co.
1853	Brig			Wm. Welt & Co.
1854	Brig	246	*E. Drummone*	Hovey, Reed & Co.
1854	Ship	1096	*Ocean Belle*	Reed, Welt & Co.
1856	Schr	185	*William Jones*	Smouse & Welt
1856	Schr	199	*Damon*	Reed & Welt
1859	Ship	657	*Village Belle*	Reed, Welt & Co.
1860	Schr	199	*G. W. Rawley*	"
1860	Ship	1100	*Weston Merritt*	"
1861	Schr	231	*Sedona*	"
1862	Schr	178	*White Sea*	"
1863	Schr	350	*Stephen G. Hart*	?
1864	Ship		*Sarah C. Welt*	Reed, Welt & Co.
1865	Schr	265	*Grae Clifton*	Benner & Welt
1865	Bark	685	*Joseph A. Davis*	Reed & Welt
1866	Schr	254	*Jesse Hart 2nd*	Reed, Welt & Co.
1866	Half Brig	277	*Mary W. Huppen*	Welt, Reed & Feyler
1866	Schr	276	*Starlight*	Reed, Welt & Co.
1867	Bark	635	*Mary G. Reed*	"
1867	Bark	611	*Rosetta McNeil*	A.R. Reed & Co.
1868	Ship	1496	*Annie Fish*	Reed, Welt & Co.
1869	Ship	1122	*Alex McNeil*	"
1871	Schr	210	*Samuel Hart*	"
1871	Schr	225	*Stephen G. Hart*	"
1872	Schr	280	*Clara G. Loud*	"

1872	Schr	237	*Nathan A. Farwell*	Reed & Caldwell
1873	Bark	873	*Alice Reed*	A.R. Reed & Co.
1874	Ship	1435	*Rosie Welt*	Reed, Welt & Co.
1875	Schr	395	*John W. Welt*	A.R. Reed
1875	Ship	1550	*Isaac Reed*	A.R. Reed & Co.
1876	Bark	747	*Annie Reed*	Reed, Welt & Co.
1877	Ship	1449	*Willie Reed*	A.R. Reed & Co.
1879	Bktn	620	*Stephen G. Hart*	Welt, Caldwell & Co.
1880	Ship	1564	*Emily Reed*	A.R. Reed & Co.
1881	Schr	425	*Fannie L. Child*	Augustus Welt & Co.
1883	Schr	488	*Maggie G. Hart*	"
1884	Ship	1745	*George Curtis*	A.R. Reed & Co.
1885	Schr	645	*Maggie S. Hart*	"
1886	Schr	852	*Josiah Hart*	George Welt & Co.
1889	Schr	1162	*Augustus Welt*	Welt & Co.
1892	Schr	777	*Ida S. Southard*	A.R. Reed & Co.
1892	Schr	750	*Madalene Cooney*	"
1900	Schr	2075	*Fannie Palmer*	George L. Welt
1901	Schr	2240	*Baker Palmer*	"
1902	Schr	1763	*Paul Palmer*	"
1903	Schr	2315	*Dorothy Palmer*	"
1904	Schr	2357	*Singleton Palmer*	"
1904	Schr	2400	*Harwood Palmer*	"

Reed Genealogy:

Isaac G. (Colonel) – 1784 – 1847 (lived on Jefferson Street)
Jane – 1780 – 1856

Isaac (Honorable) – 1809 – 1887 (lived on Main Street) – son of Isaac G.
Caroline S. – 1819 – 1843 (Redington family)
Lydia E. – 1828 – 1898 (McDonald family)

George Smouse – Stepson of Isaac G.

1823-44; 1811-81; 1817-1905; 1821-74; 1815-81
Mary G., Jane A., William G., Gardner K., Edward – sons and daughters of Isaac G.

Asa Redington Reed – 1842 – 1919 – son of Hon. Isaac and Caroline (bachelor)
Susan – 1856 – 1863 – daughter of Hon. Isaac and Lydia
Alice M. – 1862 – 1914 – daughter of Hon. Isaac and Lydia (single woman)
Caroline R. – 1855 – daughter of Hon. Isaac and Lydia (six months old)
William G. – 1858 – 1914 – son of Hon. Isaac and Lydia
Emily – daughter of Hon. Isaac and Lydia
Mary G. – 1860 – 1941 – daughter of Hon. Isaac and Lydia
Another daughter ? – Mrs. Thomas Leigh ??

Reed and Welt Build a Vessel

On March 28, 1874, the busy Waldoboro shipyard of Isaac Reed and Augustus Welt launched a "beautiful 3 masted schooner, Eva C. Yates – to be commanded by Philip D. Yates of Round Pond and intended for the general freighting business." No sooner had the launching ways (located at the present day Town Landing) been cleared than workers began preparing for the next project – "stretching the keel for a ship to be 1400 tons burden." This new vessel was to be named the *Rosie Welt*, completed and in the bosom of the Medomak 7 months later – October 30, 1874. No sooner completed was the Welt and before sunset, ANOTHER keel was laid – this time for another 3 masted schooner, the *John W. Welt*. A Jonny-on-the-spot *The Lincoln County News* reporter's comment: "No grass grows in THAT yard!"

Behind the scenes owners were working diligently to bring business and jobs to Waldoboro. To begin with, investors had to be found for each project. These people – from town and from "away" – bought shares in the vessel – usually in 64ths. (The captain and the owners themselves often owned large blocks of these shares.) Once money was rounded up, timber and supplies could be ordered. Schooners would begin to be seen on the river and at the town piers each week bringing wood – very often from places in Georgia, the Carolinas and Virginia. Owners and foremen would survey all the timber for quality before it was used. Teamsters with their horses and oxen would "snake" the heavier pieces from the shore to the saws and, if supples had come by railroad, all the way from the station on Depot Street.

In the Reed and Welt yard it was Mr. Augustus Welt himself who was the General Superintendent. Under the G.S. were several so-called "Master Workmen": Gorham Feyler, Soloman Shuman, and J. William Welt. Each specialized job had a Foreman: S.O. Waltz for the joiners, Andrew Sides the blacksmiths, Alvin Newbert the painters, John Longley (of Thomaston) the riggers, Haupt and Kuhn the fasteners, and Miles Standish the caulkers.

The building of a ship, big or small, was made up of four stages: Shipwrights first fashioned the skeleton – keel (or backbone), stem (front) and stern (back), and frames (ribs). Then the fasteners and caulkers closed in the skeleton with outside planking (skin), inside planking ("ceiling"), and oakum caulking (hemp and tar) along with molten pine tar sealing the top of each crack between the planks. Sparmakers turned out the masts and yards; riggers raised them and then attached the miles of rope to hold everything in place. Finally the vessel was outfitted: Sails from a sail loft, cabinetry from the joiners, iron work from the blacksmith shop, paint, small boats, and ballast. Cargo to make the ship pay for herself and provisions for passengers and crew came last.

While the 100 or more men worked on the ship, Mr. Reed had protected the yard and its employees through a marine insurance company. This company also inspected all aspects of the job and decided on rates. Another agency had been paid to find sailors for the first voyage; the men had arrived by train and were quartered waiting for orders as the job neared its completion.

Launching day saw the hired master do his job flawlessly. The 1435 ton *Rosie Welt* went overboard Monday, October 30, 1874, sat on the flats waiting for the tide and then was tied up at Storer's Wharf. Mr. Reed had hired the tug *Resolute* of Bath to tow her (backwards!) down river; she waited at Round Pond for a few days. Then she sailed with Captain Orlando Welt and crew to New York where she loaded for San Francisco.

The project had been a success! French Lloyd's survey have given *Rosie Welt* the highest maritime rating of 3-3, 1.1 * for 11 years. The owners, foremen, and workers felt good. However, there was no time to exult. By nightfall the NEW schooner's keel lay ready on the ways and shiploads of new southern timber, ordered months ago by Reed and Welt were arriving. Grass was NOT growing in this yard! These were good times in Waldoboro.

The Lincoln County News – November 22, 1877 – "Mr. Augustus Welt and daughter-in-law Mrs. Orlando Welt will leave today for New Orleans to meet the ship *Rosie Welt* and her skipper, Captain Orlando Welt."

William Fish, Energetic Businessman

[The Waldoboro Public Library's acquistion of the large white house and barn across from the Post Office prompts the following story.] According to an 1857 map this property was owned and occupied by one William Fish, his wife, and their five children. William Fish was a very prominent and civic minded man, involved with shipbuilding and commerce most of his life (1817–1884).

Mr. Fish brought his young wife Sarah (McIntyre) whom he had married in 1841, to Waldoboro Village from his father's farm in the eastern part of town. Father (William also) had moved to these parts in 1780 after having been the skipper of a coasting vessel. One brother, Church, remained at home; two others, Abel and Samuel removed to Hope and Thomaston respectively, while young William was settling in Waldoboro.

William quickly began to demonstrate to the town a restless energy in making a success of himself. Beginning as a storekeeper and then later as a shipbuilder, he established himself as a leader in the community which at that time was enjoying a period of tremendous growth and financial success. Unfortunately, all this prosperity was dealt a crippling blow when the Great Fire of 1854 destroyed much of the village. Undaunted, citizens quickly rebuilt and improved it thanks in great part to the unwavering financial assistance of Fish, Joseph Clark, Henry Kennedy and others.

In this rebuilding William Fish contributed two significant projects. One was the Fish Block, a magnificent five story brick and granite structure on Main Street hill which housed many important businesses from its completion in 1855 to its burning in 1900. The other was the digging of an underground water reservoir, 13 feet deep, 18 feet in diameter, and capable of holding 313 hogsheads of water, installed outside the Baptist Church's Pleasant Street entrance. This reservoir, whose outlines can still be seen on the surface of Pleasant Street, proved most helpful in the fighting of later fires, until hydrants were finally voted in in 1909. The so called "Reservoir Commitee" of which Mr. Fish was an important member, was to be thanked many times over.

From 1853 to 1858, Fish, appointed by President Franklin Pierce, served as the town's Postmaster from his office in the Fish Block.

By 1860 Mr. Fish's energy turned itself also to shipbuilding. In his shipyard just south and east of the Main Street bridge he built some 21 vessels between 1860 and 1884. These were small two and three masted schooners for the most part – 250 to 300 tons – many named for members of his family – *Joseph Fish*, *Frederick Fish*, *Samuel Fish*, etc. His last effort, completed just before his death, was the *S.G. Hart*, his largest at 504 tons.

As the days of building sailing ships waned, the established yards of Fish along with Clark, Reed & Welt, Kennedy and Storer kept employment high for a time.

Unlike some of the other shipbuilders of the day, Mr. Fish does not appear to be publicity minded. Thus we have no photographs of any of his vessels. (Beginning in 1874 many Waldoboro ship photos are available.) Records of Fish ships and their stories are equally scarce. His handsome business block – the "most impressive in all of Lincoln County" is long gone, but fortunately remains in photographs for us to admire.

William Fish died at age 67 in January 1884. Sarah, his wife, had died in 1875; three of their children were also deceased. Eventually the house passed out of family hands and was known for many years as the "Will Achorn place". More recently, like many large houses, it has been used for apartments.

(Information for this article was obtained from *The Lincoln County News* and from Stahl's and Miller's Histories of Waldoboro.)

Vessels of William Fish

YEAR	RIG	TONNAGE	NAME	BUILDER
1848	Schr			William Fish
1860	“	257	*Joseph W. Fish*	“
1860	“	239	*Frederick Fish*	“
1861	“	213	*Samuel Fish*	“
1862	“	274	*Union Flag*	“
1863	“	287	*Sarah Fish*	“
1864	“	293	*Charlotte Fish*	“
1865	“	243	*Ella Fish*	“
1865	“	296	*Fannie K. Shaw*	“
1866	“	222	?	“
1867	“	68	*Agnes I. Grace*	“
1867	“	165	*William Penn*	“
1867	“	279	*Thomas Fish*	“
1868	“			“
1869	“	186	*Joseph Fish*	“
1869	“	316	*Loretta Fish*	“
1873	“	376	*Agnes I. Grace*	“
1874	Bktn	469	*Fred Eugene*	“
1881	Schr	450	*Jennie L. Hall*	“
1882	“	474	*Stephen G. Loud*	“
1883	“	504	*S.G. Hart*	“

William Fish – 1817 – 1884 (grandson of Captain William, skipper of coaster, moved to East Waldoboro 1745; son of William. Other children – Church, Abel, Samuel.)
Sarah (McIntyre) – married William in 1841; moved to Waldoboro village 1842.

William and Sarah brought up five children (all of whom predeceased them) on Main Street.

Matthews – Fish Block – Main Street – burned 1900 and replaced by the Gay Block.
The Fish Block was also known as the Union Block.
Courtesy Waldoborough Historical Society.

Chapter 2
Life At Sea

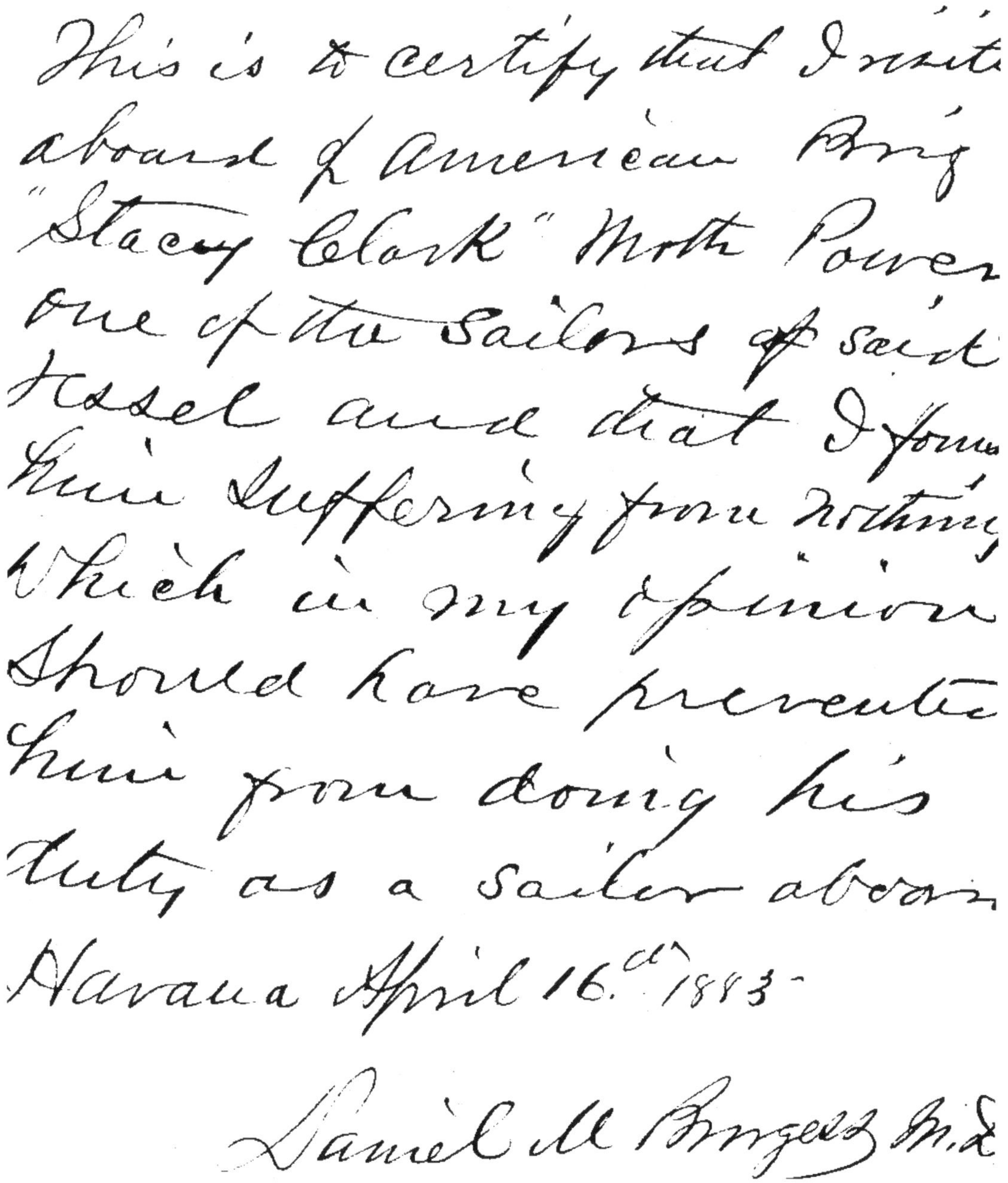

This is to certify that I visited
aboard of American Brig
"Stacy Clark" Moth Powers
one of the sailors of said
vessel and that I found
him suffering from nothing
which in my opinion
should have prevented
him from doing his
duty as a sailor aboard
Havana April 16th 1883

Daniel M Burgess M.D.

A doctor's letter concerning Seaman Powers of the Stacy Clark
Courtesy of Mrs. Herbert Sperry.

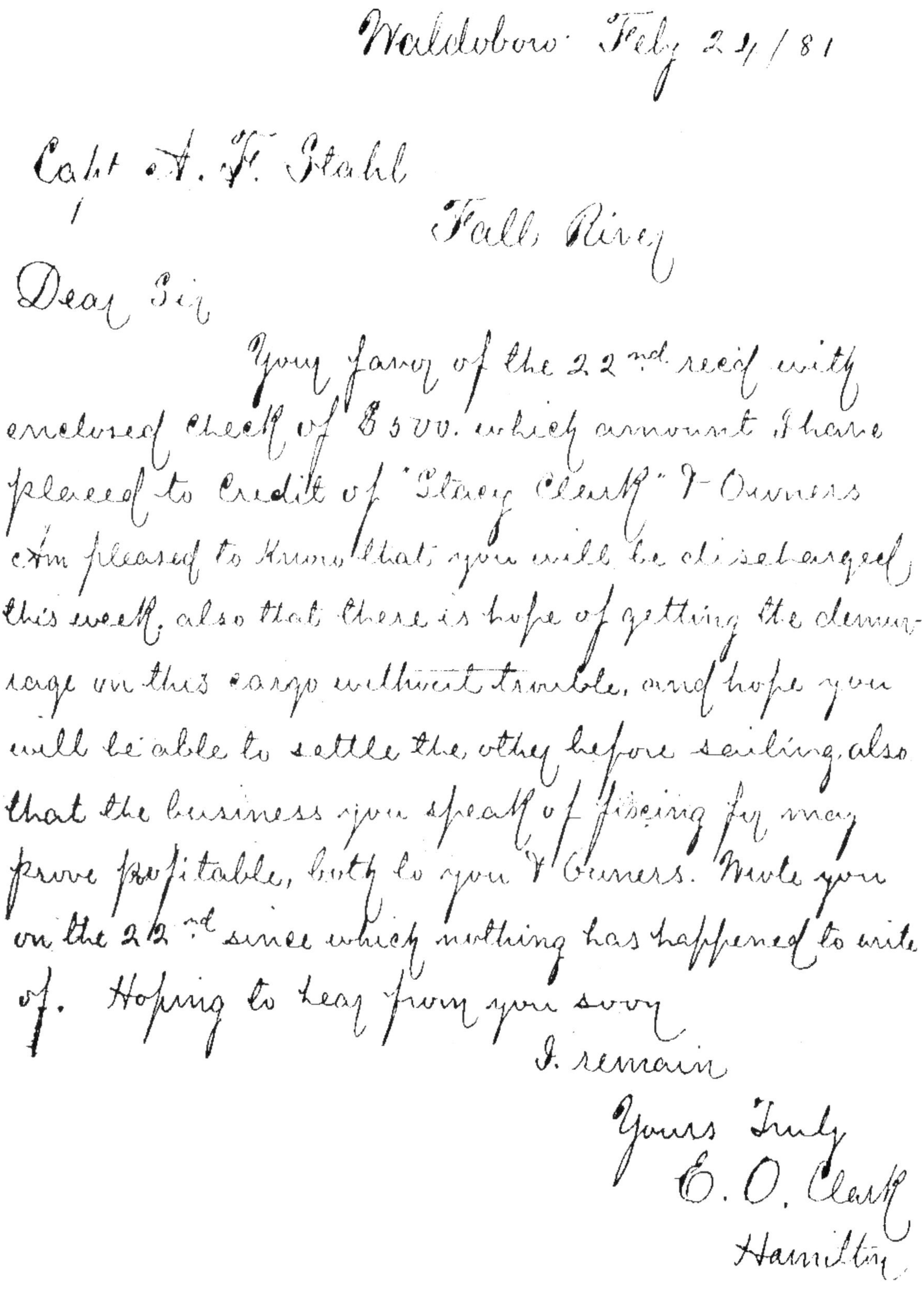

Waldoboro Feby 24/81

Capt A. F. Stahl

Fall River

Dear Sir

Your favor of the 22nd. rec'd with enclosed check of $500. which amount I have placed to Credit of "Stacy Clark" & Owners. Am pleased to know that you will be discharged this week, also that there is hope of getting the demurrage on this cargo without trouble, and hope you will be able to settle the other before sailing, also that the business you speak of fixing for may prove profitable, both to you & Owners. Wrote you on the 22nd since which nothing has happened to write of. Hoping to hear from you soon

I remain

Yours Truly

E. O. Clark

Hamilton

Letters to Captain Stahl of the Stacy Clark *from owner and builder E. O. Clark*
Courtesy of Mrs. Herbert Sperry

Waldoboro: March 11/81

Capt A. F. Stahl

Belfast

Dear Sir

Yours of the 10th rec'd with enclosed Check for $83.21 which I have placed to credit of Brig Stacy Clark & Owners. Mr Clark is still in New York and probably won't be home until after you sail. It's raining here now but hope it will clear away soon so that you can get away early next week. The "Jos Clark" arrived at the Vineyard the day you sailed & left there for Portland on the 8th. Suffered no damage coming north but splitting jib & losing some waste boards. Capt Whitney writes she is very tight now. Also notice the Josephine's arrival at Cardenas on the 27th ult. Hoping you will have a fair passage & good dispatch at Savannah I remain Yours Truly

O. O. Clark, Hamilton

Remains of the Kennedy (above) and Reed-Welt Shipyards about 1911. Author's collection

Adventures on Two of Henry Kennedy's Vessels

The Kennedy shipyard was located on the east side of the Medomak River, just south of Storer's Lumber Yard and below the old Waldoboro High School (now known as Friendship Street School). Henry Kennedy, a cousin of Joseph Clark, began the business in 1837 and built approximately 25 vessels until his death in 1875. His sons Henry and L.L. continued the H. Kennedy yard and constructed six more craft until closing down shortly after 1883. Our story concerns two of the last three schooners built by the Kennedys; it describes the risks shared by men, women, and children who went to sea.

The *Annie B. Hoffses*, 428 tons, commanded by David Hoffses and named for his young daughter, was sailing in Caribbean waters not long after her launching in 1882. Aboard were father, mother, daughter Annie and son Lincoln. Captain Hoffses was a commander of remarkable skill who had never in his career had even the smallest accident while skippering ships. In a Cuban port he was taken ill and had to retreat to his bed. His family, homesick for Waldoboro, yearned to depart and persuaded father to turn the running of the schooner over to his son. The *Annie B. Hoffses* with her family aboard, sailed for Maine – and never was heard from again.

1883 saw the 407 ton schooner *Ida Francis* launched by the Kennedy brothers. Her captain was Fernando Francis; the vessel was named for his wife, a Friendship girl who, with their infant son, was accompanying her husband. Loaded with lumber, the *Francis* headed for sea – and, as it turned out, dreadful danger. Before long a hatch cover was discovered to have worked itself loose; the ensuing leak caused the heavily loaded ship to sink so that only part of her stern remained above the water. In the lazarette – a tiny stern compartment – the family and crew huddled for safety and shelter and for six weeks endured almost indescribable suffering and hardship. All food was gone (even the pet parrot – carefully saved as long as possible – was eventually eaten.) With hope lost, suddenly a miracle! A cry went up that the sails of another vessel had been spied, and the starving and emaciated group was soon to be rescued from their fate.

Strangely, it appeared that the skipper of the rescuing vessel was a close friend of Captain Francis. For several days this man had been bothered by an unaccountable anxious feeling concerning Francis' whereabouts. So the little family was made safe and brought home to Waldoboro. But in later years when the father attempted to relate the tale, he would break down utterly when he recalled the terrible suffering he had unwittingly brought upon his wife and son.

Schooner Ida Francis – *"In a tiny stern compartment the family and crew huddled for safety."*
Sketch courtesy of Thomas D. Turlo

J. Manchester Haynes, *Alfred Storer – 1885 loading lumber at St. Simons, July 1902. Courtesy Maine Maritime Museum*

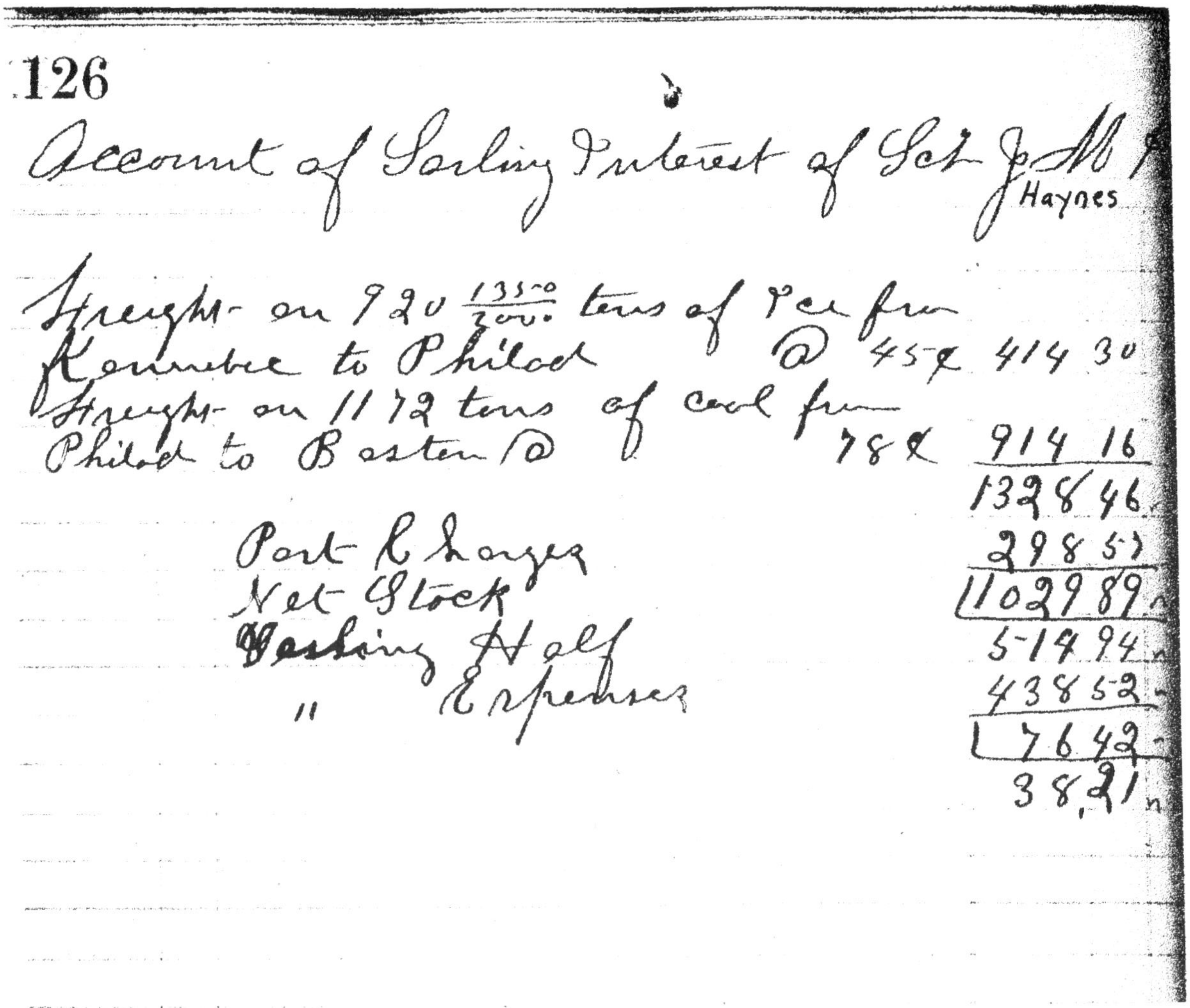

126

Account of Sailing Interest of Sch J M Haynes

Freight on 920 1350/2000 tons of Ice from Kennebec to Philad	@ 45¢	414 30
Freight on 1172 tons of coal from Philad to Boston	@ 78¢	914 16
		1328 46
Port Charges		298 57
Net Stock		1029 89
Sailing Half		514 94
" Expenses		438 52
		76 42
		38 21

Two pages copied from the account book of Captain Jonathan Matthews of the schooner *J. Manchester Haynes*. The book covered Matthews' voyages in the *Haynes* from 1885 to 1897. Most cargoes were the same – ice from the Kennebec south to Baltimore or Philadelphia – and coal on the return trip– usually to Boston or New York. Sailing expenses – including wages for captain and crew are also noted.

Captain Matthews eventually married one of Levitt Storer's sisters and settled in Waldoboro.

127

No 61 from Aug 24 up to Sept 30 1883

Sailing Expenses

	To 34 lbs of Beef		2 07
	" 10 Lamb & Sweet Potatoes		1 15
	" Meat		1 35
	" Tomatoes 20 & Clams 40		60
	" Dispatch		29
	" George N Johnson		12 29
	" Shipping Com Man		1 00
	" George Martinson	Wages	4 17
Sept 19	" Horace House	"	22 50
" "	" John Allen	"	22 50
" "	" Peter Anderson	"	22 92
" "	" Harry Jackson	"	18 33
" 13	" E & J Baker	Bill	11 58
Sept 29	" J Baker & Co	"	55 91
" "	" H H Baxter	"	4 00
" "	" H Nelson	Wages	55 50
" "	" Willard Nickerson	"	55 50
" "	" Charles Franklin	"	43 17
	Marketing		5 02
	" Jonathan Matthews	Wages	98 67
			438 52

From the collection of Captains Douglas and Linda Lee

J. Manchester Haynes *rounds Storer's Point on her way out of Waldoboro – 1885. Courtesy Penobscot Marine Museum, Searsport.*

Edwin Achorn and his Clipper Ships

From the thousands of wooden sailing ships built in the United States, the term "clipper" evokes the greatest feelings of excitement, pride, and patriotism. Carl Cutler, in his definitive account, *Greyhounds of the Sea*, describes a "clipper" as being a vessel 1) full rigged, with a concave bow, and very sharply built; 2) extremely heavly sparred; 3) commanded by a captain who spared no effort to drive her hard to extract every bit of speed out of her.

Although some smaller vessels of this type had been built before 1849, it was the discovery of gold in California and Australia and the subsequent rush to reach these places that spurred the clipper building era. This era was to last only until 1857 when an economic depression or "panic" ended it. But it was a glorious era indeed with names like Donald McKay, *Flying Cloud*, and *Sovereign of the Seas* remaining etched forever in history books. Maine builders from Rockland, Damariscotta/Newcastle, and Bath contributed also to the lists of "sharp" ships.

Waldoboro builders seem to have shied away from the newfangled ideas, preferring a more conservative approach in their style. They built ships extensively – in excess of 200 vessels between 1849 and 1857, by the records we have, but only four of these were classified as clipper ships. The primary clipper builder in town was Edwin Achorn (1809 – 1887) whose yard was located just south of the current Town Landing. The Achorn family constructed at least 24 vessels with Edwin involved in 18 – including the four clippers.

1852 saw two of these sharp ships enter the Medomak. The first was the *Woodcock* – 1091 tons, 187 feet long and 35.4 feet wide. Cutler indicates that she enjoyed a good, unspectacular career and eventually was wrecked at Dungeness, Ireland. The captain at the time of her demise was Mr. Lambert.

Wings of the Morning also slid down Edwin Achorn's ways in 1852. She was owned by Talbot & Olyphant of New York for several years in the Australian trade. After one relatively slow run of 102 days, New York to Melbourne, her time was sarcastically commented upon by a writer in a maritime magazine: "The *Wings of the Morning* came in day before yesterday from New York, but the *Uttermost Parts of the Sea* has not been heard from. *Snail* or *Tortoise* or *Drone* are possible names for the next clipper – I am tired of these always-a-little-faster-names!"

Wings, however, did have some good sailing days. On a run from San Francisco to the Pacific equator she finished in 15 days in early 1856, and she beat the swift clipper *Red Rover* by 19 days on a run from Callao to Havre in the same year. On two west bound voyages in 1853 and 1855 around Cape Horn to San Francisco she completed the journeys in 165 and 145 days. In an earlier voyage she was partially dismasted in a storm and had to put in to Rio De Janeiro.

In 1856 she was sold to the French and renamed *Surrat*. She was still registered in 1868, owned in New York by Dunham and Dimon and skippered by Frederick M. Ranlett. Prior to her last listing she established an all time record for sailing time between San Francisco and Shanghai, China.

In 1853 the third Achorn clipper was launched. This was called *Spark of the Ocean* – 171 feet long, 33.7 feet wide, 16.9 feet deep with 895 tons carrying capacity. According to Cutler, she was owned by Alfred Blanchard and William Greeley of Boston and was commanded by Willard Treat.

The last of Edwin Achorn's clippers was the largest and, as it turned out, the most tragic. This was the *Achorn* of 1854 – 1250 tons. She made but one short voyage after being launched. Her end is described by J.J. Stahl in his *History of Old Broad Bay and Waldoboro*, Volume 2:

> "In order to get out of the Medomak before the river closed for the winter, she proceeded to Muscongus (a common occurrence) there to complete her fitting out. But on the morning of January 3, 1855, in some unaccountable manner, the *Achorn* took fire and was completely destroyed."

Fortunately for the major owners – out of town parties – the ship was heavily insured, and they recovered their investments. For the smaller shareholders, including some Waldoboro folks, it was a different story. Alexander Young, the master builder, was forced to mortgage his fine brick village home to cover his losses. Edwin Achorn himself seems to have lost his fortune and his nerve with the destruction of the *Achorn*. Not until 1871 did he return to shipbuilding. With his son he finished his career with four small schooners ending in 1874.

A footnote to the sad story of the *Achorn* came from Philip Achorn, a descendant of one of Edwin's brothers. After the fire on January 3, 1855, the ship sank into the waters off Muscongus somewhere between Ram Island and Hockomock Point. Today she is said to remain on the bottom in this area; lobster fishermen have occasionally entangled their lines in her rigging.

A true clipper – Black Warrior – *built in Newcastle, Maine in 1853 by Algernon Austin. Courtesy Peabody-Essex Museum*

Advertised as a true clipper – Ellen Austin – *built in Newcastle in 1854 by Nathaniel Austin. Courtesy Peabody – Essex Museum*

The Bark *Saint Cloud* and War with Mexico

(Author's Note – The fascinating road of local historical research often runs in far reaching circles: Not long ago I was contacted by a Captain Paul R. Rosewitz of the U.S. Army, an Assistant Professor of Military Science, Kemper Military Junior College, Boonville, Missouri. He was seeking information about a three masted bark *Saint Cloud* built in Waldoboro and used, he said, as a military transport in the Mexican War. It is from our subsequent correspondence and research that the following is derived):

On May 11, 1846 President James K. Polk sent a war message to the United States Congress. "The cup of forbearance has been exhausted," he declared. "After reiterated menaces, Mexico has passed the boundary of the United States, has invaded our territory, and has spilled American blood upon American soil." Congress soon responded by declaring that a state of war now existed between the two countries.

(Some words of explanation: Polk, elected largely for his promises of westward expansion, had wanted for some time to gain control of California, a Mexican province, Oregon, another Mexican province, and Texas, originally Mexican but now having broken free by civil war, an independent state. California, with its forests of pines and giant sequoias, broad wheat filled valleys, vineyards, and mineral deposits (no gold yet!), was particularly attractive to the President. Barely 6,000 white people lived there, the native Indians were not to be feared, and the Mexican overlords feeble. It seemed ripe for the picking, and the popular opinion, expecially from southern and western Americans, cried out for it. Polk offered to buy the territory; Mexico refused and furthermore commenced protestations about America's urgings and help to Texas in her civil war. Texas then formally joined the United States of America, and the president sent forces there to "protect the new citizens". When shooting broke out between Mexicans and the U.S. army, Polk saw his opportunity to keep his promises about expansion of the country, and to checkmate any dreams of rival empire builders Great Britain and France of controlling Pacific settlements. The war was on with great enthusiasm from many but not all Americans.)

Southerners and westerners sent 49,000 volunteers; the original 13 colonies were much less forthcoming with troops – about 1,000 from each. (Waldoboro supplied one man, William H. Stahl, to the fray.) Although European sages doubted that the untrained, "soft" Americans could stand up to the trained, "wiry and tough" Mexicans, the struggle was over relatively quickly, and the Treaty of Guadeloupe – Hidalgo gave the U.S. decisive victory and huge additions to its continental ownings. California and Texas gave a sea to shining sea look to a new America. Mexico received $15,000,000.

William H. Stahl, Waldoboro's lone addition to the army, was sent to be part of the very first amphibious U.S. invasion in its history. The landing was made on a beach at Vera Cruz, Mexico, a port not far east from the capital, Mexcio City. 12,000 troops, under the command of General Winfield Scott came ashore efficiently with horses, artillery, and supplies in 24 hours with the aid of "surf boats." Soon Vera Cruz was under Scott's control and he was setting forth for Mexico City and eventual overall victory.

One of the vessels that made the journey to Vera Cruz was the bark *Saint Cloud*, 100.4 feet long, 24.5 feet wide, and 13.1 feet in depth. The *Saint Cloud* was built in either (we are not sure which)

1841 or 1844 by Edwin Achorn, a Waldoboro builder who constructed at least 12 ships (including 3 clippers) between 1836 and 1854 at his yard just south of the present Town Landing. In the beginning of her career she was commanded by Francis Davis, and she was owned at that time by her captain, her builder, and by Rufus Achorn, Andrew Sides, and Harrison Woodbridge. In 1847, she was commissioned by the U.S. Army to carry 110 members of the Mountain Howitzer and Rocket Battery, their ammunition stores along with the 2nd Artillery headquarters AND its band from Fort Monroe, Virginia to Vera Cruz. (Author's note – seemingly a tremendous load for a 100 foot vessel despite her Waldoboro craftsmanship! Small wonder that the return voyage was on another ship; unfortunately, the Mountain Howitzer and Rocket Battery only returned 60 of its 110 men.)

It is most interesting to note that, despite an apparent lack of enthusiasm by its inhabitants for the war with Mexico, we all can be proud of the little Waldoboro bark *Saint Cloud*. She played an important part in a great American victory – a victory which abruptly changed forever the boundaries of our country and its influence in the world.

(Sources for this article include Jasper Stahl's *History of Old Broad Bay and Waldoboro*; Samuel Eliot Morison's *Oxford History of the American People*; The writings of Robert Applebee, Marine historian, and the research of Captain Paul R. Rosewitz, U.S. Army.)

Landing of the troops at Vera Cruz

Down Easter Emily Reed – *built in 1880 by A.R. Reed & Co. – 1564 tons. Author's Collection*

The Down Easter *Emily Reed*

The Waldoboro Town Landing today is primarily used as a launching ramp for the 40 or 50 clam diggers who make their living from the river and other coastal regions. Historically, however, the town landing was the building site for some of the largest ships ever launched from this village. Here was where the famous Palmer schooners – 6 of them – were constructed between 1900 and 1904 by George Welt. Prior to that, it was the shipyard of the Reed family producing over 50 schooners and square riggers between 1837 and 1892.

In the hot summer of 1880 over 100 men were in the employment of Asa Redington Reed who by 1867 had taken over control of the family yard from his father, Isaac Reed. Although work had had to be suspended twice during June and July – once because of a lumber shipment failure and once because of most employees being excused during haying season – the vessel they were building was progressing well. Mr. Reed announced to *The Lincoln County News* October 14 that a November launch was expected. He said also that Thomaston sail loft, George Washburn & Sons, was preparing suits of sails.

Finally, on November 10, the large vessel (1600 tons, 209' long X 40'6" wide X 24' deep) slid successfully into the cold Medomak. Several feet longer than any ship heretofore launched at Waldoboro, she was to be called *Emily Reed* (for one of the builder's sisters). Her captain was O.D. Sheldon of New Jersey; her managing owners were Yates and Porterfield of New York. A square rigged ship or "Down Easter," she was sturdier than most clippers and nearly as fast. As was customary after a Waldoboro launching, a tug towed her part way down river where she anchored off Hollis Point. From here she was scheduled to sail to New York where she was to load case oil for Calcutta, India.

During her 28 year career, the *Emily Reed* was described as a "good ship and good carrier which made fair passages". But in her world wide voyages after leaving Waldoboro (and never returning) she encountered several interesting mishaps – somewhat unusual even for the rigorous business she had entered:

1885 – Experiences severe weather off Cape Horn during which the cargo shifted, a number of spars sprung, and most of her fresh water lost.

1888 – Had to put back into Valparaiso for repairs, reconditioning, and restorage of part of her cargo.

1890 – Left San Francisco in July with a cargo of wheat for Rio De Janeiro. Another disaster off Cape Horn where heavy gales and tremendous seas kept the ship practically submerged for 4 successive days. Forecastle stove in and 8 crew members swept overboard.

1891 – On return trip from New York to San Francisco, rudderhead sprung in a gale far down in South Atlantic. Had to put into Rio De Janeiro for repairs and new rudder.
After making San Francisco in 95 days, Captain O.D. Sheldon relinquished command of the vessel in disgust because of the seeming bad luck which followed him. Later she was sold by Yates and Porterfield to Hind, Rolph, and Co. of San Francisco.

Finally, in 1908, the *Emily Reed* was sailing in dense fog off the northern coast of Oregon. Her chronometers, according to her captain Kessel, showed her to be 60 miles off shore when she suddenly struck and went ashore near Nehelem Bay. The *Emily Reed* was lost February 14, 1908.

(Author's note – Information about the 28 year career of the Reed comes from William Fairburn's *Merchant Sail* Vol. V.)

A postscript to this story is an old letter found at the Waldoborough Historical Society. A resident of the Nehelem Bay area, while beachcombing, uncovered a fragment of wood. On the fragment was carved the name *Emily Reed*. A. R. Reed's sister's ship had been found.

The Incredible End of the Ship *Emily Reed*

Excerpts from the book *Oregon Shipwrecks* by Don Marshall and sent to us by George Leach of Melrose, Mass. have told us more about the ship *Emily Reed* described earlier. Readers will remember that the *Reed* was launched November 10, 1880 by Asa Redington (A.R.) Reed at the site of the present Waldoboro Town Landing. At 209 feet long and 1660 tons she was the largest ship heretofore built in Waldoboro; she was to embark upon a long career spanning 28 years, taking her to all parts of the world but never again seeing her home town.

In February of 1908 the *Reed*, now rigged as a bark, was approaching the Oregon coast with a load of coal for the hungry industries of Portland. The skipper, Captain Kessel, scanned the darkness, hoping to catch a glimpse of a comforting arm of light from the Tillamook Lighthouse. He was unaware that the *Reed's* chronometers were faulty, and that he and his companions were steering for too close to the beaches of Nehalem Bay. The February weather had turned typically dark and overcast.

Suddenly the blinded bark drove with a smashing crash into the breakers and hit the beach bow on, at full tide. Her backbone (keel) broke almost instantly under the weight of the coal; her bow section listed to port, and the men there were swept into the sea. Captain Kessel, his wife, Second Mate Charles Thompson, Seamen Barney Sullivan, H. Franchez, and Herman Bertel all clung to the sea washed poop deck until the tide waned, and they were able to scramble to safety through the surf. Because it was unseen, they assumed that the ship's lifeboat had either capsized or disintegrated in the tangle of broken spars and lines. They further assumed and later reported that the remaining 11 men had perished. Both of these assumptions later proved to be wrong.

Three days later and 200 miles to the north, the 6 ton sloop *Tecla* bobbed at anchor in Neah Bay, Washington. Her crew, hearing a weak voice calling them from starboard, was able to discern a lifeboat with men in it. It was the *Emily Reed's* life boat, and it contained Seaman Fred Zube and two other crew members. This group had managed to guide their tiny craft 200 miles with one oar; they had no water, food or medical supplies, but they were alive. Another shipmate, the *Reed's* cook, lay dead on the floorboards, having succumbed to the terrible ravages of extreme thirst.

The news of the tragedy and the miraculous journey was soon telegraphed to Portland and soon after that back to the East and Waldoboro. The *Emily Reed* was a total loss; her bones were buried deep in the sand of that beach, subsequently to be uncovered by the wind, found in 1938, and photographed by later 20th century marine historians and others.

Because of this wreck and others in the same area of the Oregon coast, a Life Saving station was established with a permanent eight man crew under the command of a Captain Farley.

SHIP EMILY REED

WRECKED OFF OREGON COAST NEAR NEHALEM RIVER AND IS TOTAL LOSS, WITH HER CARGO.

Wrecked off Oregon Coast near Nehalem River and is total loss, with her cargo.

FEBRUARY 15, 1908.

PRICE FIVE CENTS.

EMILY REED GOES ASHORE ON ROCKS

Vessel and Coal Cargo Total Loss.

STRIKES NEAR NEHALEM RIVER

Chronometers Are Wrong and Ship Gets Too Close In.

TEN OF THE CREW LOST

Captain Sights but One Vessel on the Voyage From Newcastle — Struggles Against Storms—25 Per Cent Reinsurance Offered.

SAVED.

Captain Kessel.
Mrs. Kessel.
Charles Thompson, second mate.
Barney Sullivan, seaman.
H. Franches, seaman.
Hermann Bartell, seaman.

LOST.

DUBIE, mate.
WESTLUND, carpenter.
HIRSCHFELD, cabin boy.
DIXON, seaman.
DARLING, seaman.
JAHNKE, seaman.
COHENSTAD, seaman.
GILBERT, seaman.
ABLISTEDT, seaman.
JOHNSON, seaman.

BAY CITY, Or., Feb. 14.—(Special.)—The American wooden ship Emily Reed, 103 days out from Newcastle, N. S. W., with a cargo of coal for Portland, went ashore half a mile South of the Nehalem River, on the Oregon Coast, at 1:30 o'clock this morning. The vessel immediately broke in two and ten members of the crew were drowned. Captain Kessel, his wife, Second Mate Charles Thompson and three members of the crew were saved.

The survivors were brought to Bay City yesterday morning and are now quartered at private residences. The ship has broken up and will be a total loss, as will also the cargo of 2110 tons of coal. The vessel was consigned to the Pacific Coast Company at Portland. Owing to the long passage of the ship, 25 per cent reinsurance has been offered on her.

the ship, 25 per cent reinsurance has been offered on her.

Heavy Weather Prevailing.

During the past few days heavy weather has been prevailing off the Oregon and California Coast. The chronometers of the Emily Reed were off and the master worked her too close in to shore before he discovered his error. Captain Kessel was endeavoring to make Tillamook Rock. He was correct in his latitude, but too far to the eastward. When he discovered his position it was too late to wear ship and she struck on one of the most dangerous places on the Oregon beach.

The Reed struck at 1:30 o'clock in the morning, bow on the beach. There was a heavy sea running and a strong flood tide. When she hit the beach her back broke and the forward end took a list to port. An effort was made to launch a life-boat. It was put over the side with three men in it. As soon as it hit the water it swamped and the occupants drowned before the eyes of those left on board. The men forward were swept from the deck by the waves. The mate was washed overboard and drowned while trying to direct the movements of the men forward.

Strikes at High Tide.

The Captain stuck to the poop and forced his wife to remain below. The second mate and three men in his watch were stationed on the main deck. When the forward portion of the ship listed they succeeded in making their way aft. By clinging to that portion of the ship until daylight they made their way ashore. It was then dead low water. It was fortunate that the vessel struck on the top of the high tide, or none of the officers or crew would have ever reached the shore to tell the tale of horror and suspense.

At a late hour this afternoon there is scarcely anything left of the once-famous ship. Her coal cargo is strewn along the beach for several hundred yards and it is doubtful whether any of it can be saved.

Tale of Hard Luck.

Captain Kessel, who is now with J. O. Bozarth at Bay City, said this evening in speaking of the wreck:

"My chronometers went wrong and I missed my latitude. I had endeavored to make Tillamook Rock and was correct in latitude but too far to the eastward. I had hard luck all the way from Newcastle and that surely did not desert me. When six days out from the Australian port I was spoken. That was in 22 S. 179 W. Since that time I have seen no one. For weeks I bucked head winds and heavy weather. One storm after another was encountered. This morning I was called shortly after the hour of 12 and discovered that we were too close in. I endeavored to haul out but was too close in. The heavy breakers and the flood tide carried me well up on the

(Concluded on Page 4.)

EMILY REED GOES ON ROCKS

(Continued From First Page.)

beach. It was exactly 1:30 when we struck. The forward part of the ship went down. The second mate, three seamen, my wife and myself, who were aft, succeeded in staying by the ship until low tide at daylight."

The Emily Reed was one of the old-time American ships, and was of 1466 tons net register. She was owned by Hind-Rolph & Company of San Francisco. She was built by the Emily Reed Ship Company in 1880 and was sold by them to the California firm. She is well-known in the local harbor.

The cargo on the craft consisted of 2110 tons of Australian coal and was consigned to the Pacific Coast Company at Portland. A. M. Stevens, agent of the company, has not decided as to what steps will be taken to salve the cargo.

SHIP EMILY REED

WRECKED OFF OREGON COAST NEAR NEHALEM RIVER AND IS TOTAL LOSS, WITH HER CARGO.

Captain and Wife, 2nd Mate and Three Seamen Safe.

CARGO OF COAL LIES ALONG NEHALEM BEACH

Master of Lost Ship Says His Chronometer Led Him Too Near Shore.

BAY CITY, Or., Feb. 15.—Eleven members of the crew of the Emily Reed, American ship, were drowned, and the ship herself was lost early yesterday a short distance south of the mouth of the Nehalem, off the wild coast of Northern Tillamook.

Captain Kessel started at 9 o'clock this morning to visit the scene again and see if any of the cargo could be saved. He thinks, however, that everything is lost.

Captain Kessel, his wife, Second Mate Charles Thompson and Seamen Sullivan, Bartell and Franches alone were saved. All came down to Bay City, which they reached yesterday afternoon. Mrs. Kessel is suffering from shock, but was not much hurt. The other survivors are resting here until word is received from the owners.

The dead are: First Mate Duble, Carpenter Westlund, Seamen Sortsell, Johnsen, Jahnke, Dickson, Darling, Cohenstad, Gilbert and Ablistedt, and the cabin boy, named Hirschfeld.

The Emily Reed was en route to Portland from Newcastle, N. S. W., with 2100 tons of coal for the Pacific Coast Company. She has had all kinds of trouble on the trip and was so long overdue that 25 per cent reinsurance was offered.

Captain Kessel, in a statement for The Telegram, blames his chronometers for allowing him to get too close in. "I was correct in latitude, but too close to Nehalem beach," he said. "When I was called, I was in the heavy breakers, and the tide and breakers together carried me in."

Three of the seamen were drowned trying to make shore in the lifeboat. The mate was washed overboard. When the boat struck the forward part went down and seven others were swept overboard. Captain Kessel and the others got aft and stayed there until daybreak, when tide was out and they got to shore.

The coal is scattered over Nehalem beach.

Below Tillamook a short distance the American bark Emily Reed was sighted at 2 o'clock yesterday morning by the steamer Washington, which reached Portland from San Francisco this morning. Though only about five miles off shore, she was standing in toward the breakers, making something like six knots an hour, as near as could be judged.

Her identity was not then known, but after having read an account of the wreck the officers of the Washington say they are positive in their own minds that the stranger was the Emily Reed. She was a full-rigged ship, and answers her description in every way. The Washington was within a quarter of a mile of her and hugging the coast pretty well to escape as much as possible a strong northwest wind.

"I know she was not more than five miles from the shore, and she was headed straight for it," said First Officer Avery. "There did not appear to be anything wrong with her, but we thought it rather strange that she was proceeding in that direction, and remarked upon it among ourselves. It was raining and the weather was thick. She had no signals flying. It is possible that she did not see us, although we were close. We were a little too far away, however, to speak her."

Robert Jones, cook on the Washington, served in a similar capacity on the Emily Reed less than a year ago. His last trip on her was from Puget Sound to San Pedro.

"During that voyage," said Jones this morning, "the members of the crew were kept busy at the pumps to keep the old ship afloat. She was leaking like a sieve. In the galley I was standing in water which reached to my knees during the greater part of the trip. There seemed to be something wrong with her all the time. The truth of the matter is that she was about worn out on account of her age. Whenever she reached a port she had to be repaired."

The Emily Reed was well known at Portland, where she was a frequent visitor. She was a sister ship of the Isaac Reed, which has also taken a number of cargoes from the Willamette River to California ports. The wrecked craft be-

EMILY REED LOST

(Continued from First Page.)

longed to Hind, Rolph & Co., of San Francisco. She was 215 feet long, 40.6 feet beam and 24.1 feet depth of hold, being of 1408 net register tons. She was built at Waldsboro, Me., in 1880, and was one of the few old American wooden ships in service out of Pacific Coast ports.

It was November 1 that the Emily Reed cleared from Newcastle, N. S. W., for Portland, with a cargo of 2100 tons of coal, consigned to the Pacific Coast Company. As the cargo belongs to the owners of the ship until delivered, it is scarcely likely that the captain will be obliged to make a business trip to Portland in connection with the matter. So far the local agent of the Pacific Coast Company has received no word from him.

The ship was about two months overdue.

EMILY REED A TOTAL LOSS

SHIP TOTAL LOSS

Shore Strewn for Miles With Wreckage of Emily Reed.

CARGO CARRIED OUT TO SEA

Hardly Vestige of 2110 Tons of Coal Visible on Beach—Total List of Lost, Eleven—Survivors Are Comfortably Cared For.

BAY CITY, Or., Feb. 15.—(Special.)—Wreckage from the ill-fated American ship Emily Reed, which went ashore near the mouth of the Nehalem River at an early hour yesterday morning, lines the beach for two miles to the southward of the place where the vessel struck. The cargo of coal which the ship carried has been washed out to sea and hardly a trace of the 2110 tons which she had on board is visible. Both ship and cargo are a total loss.

The total list of lost from the Emily Reed numbers 11. The name of the seaman which was not forwarded in the dispatch of yesterday is not known. He was a Russian and none of those saved

DEAD REMAIN IN GRASP OF OCEAN

Search Fails to Reveal Victims of the Lost Ship Emily Reed.

WRECK BREAKS UP AND DRIFTS ASHORE

Captain Kessel Patrols Beach in the Vicinity of Bay City in Hopes of Finding Bodies of Crew Washed Upon the Sands.

Captain Kessel of the wrecked American ship Emily Reed patrolled the beach yesterday afternoon for a long distance opposite where the vessel went to pieces early Friday morning in hopes of finding some of the bodies of the 11 men who lost their lives. His search proved in vain.

Large quantities of wreckage are drifting ashore and the coal cargo is washing up on the sands. Little will remain of it in the course of a few days, owing to the action of the waves, and there is now no hope of saving much of it. All told the ship carried 2,100 tons.

The Emily Reed went hard upon the beach, and broke into splinters almost at once. Captain Kessel had lost his reckoning and thought he was quite a distance to sea when suddenly the shore line of Tillamook bay loomed up before him. The officer in charge had called him, but it was too late to alter the vessel's course and all that could be done was to attempt to save as many lives as possible. Seven of the men went down with the forward part of the ship when it crumbled away from the force of the contact with the bottom. Three were lost by the swamping of the life-boat, and the chief officer was washed overboard on the crest of an immense breaker. All could have been saved had they rushed for the after part of the vessel where the captain, his wife and four of the crew sought refuge. They remained there until low tide when it was possible to reach shore without much trouble.

Captain Kessel will remain at Bay City until all hope of finding the bodies of those lost has been abandoned.

THE EVENING TELE

VOL. LXII. CITY EDITION. PORTLAND, OREGON, MONDAY, FEBRUARY 17, 1908.

THREE MORE SAVED FROM EMILY REE

THEIR LIFEBOAT NOT SWAMPED

Rowing 200 Miles N. They Get to Shore at Neah Bay.

COOK ALSO IN BOAT, BUT DIES FROM EXPOSURE

Dubie, First Mate, and Two Sailors Taken From Death List.

(Telegram [illegible]cial.)

PORT TOWNSEND, Wash., Feb 17.—First Mate Dubie, Seaman Arthur Jantuke and Seaman Eauld Abelstedt and the dead body of the cook, of the wrecked wooden ship Emily Reed, came ashore last night near Neah Bay, on the Straits, a short distance from Tatoosh lighthouse.

This unexpected survival of three of the man who have been counted among the lost since Friday morning reduces the Emily Reed death roll to eight. The cook was alive until yesterday, when he died from exposure and lack of food.

These survivors put off from the Emily Reed shortly before the ship grounded, and their open lifeboat was reported swamped by Captain Kessel. Instead of being drowned, however they managed to get away through the darkness and the storm and started rowing north.

It is nearly 200 miles from Nehalem beach to Neah Bay, and these men made the trip in the small boat with-

(Concluded on Page 2.)

THREE MORE SAVED.

(Continued from First Page.)

out water, provisions or rest. The three remaining alive are exhausted, but will recover.

They were almost blown ashore off the Washington coast.

The revenue cutter, which is stationed here to render quick aid to mariners in distress along the coast of Washington, is steaming up for the run to Neah Bay.

SEATTLE, Wash., Feb 17.—A dispatch from Neah Bay says:

The cook died in the boat at 3 A. M. on the 16th of exposure. As the men were only partly clothed, it was very hard for them to endure the exposure.

Mate Dubie received a broken arm by the wreckage. The survivors will probably all recover. Dr. Wood is attending them, but they are all very weak.

Today the men received the first food and water they had had since Thursday night.

Half a mile south of the mouth of the Nehalem River, the point at which the Emily Reed was wrecked, to Tatoosh Island is approximately 175 miles in a direct route. The course taken by the small boat containing the survivors would zigzag more or less, owing to the vagaries of the wind, which would lengthen the distance actually covered to more than 200 miles.

During the past four days the general trend of the wind has been from the southeast and would have carried the frail craft north and far offshore had she not been properly managed. Off the Oregon coast Friday and Saturday the wind was sweeping along in the course indicated at from 12 to 52 miles an hour, according to the official records. It had a mean velocity of about 26 miles an hour most of the time.

Besides, there is a strong current running north, which does not set in toward shore until well up along the Washington coast. Gas buoys getting adrift at the mouth of the Columbia River have shown the speed of the current to be little short of phenomenal. A couple of years ago a buoy drifted from Grays Harbor to Vancouver Island in less than three days.

Heavy swells were running and it would have been equivalent to suicide for the occupants of the boat to attempt to reach shore along the Oregon or Washington coasts. Their craft would have been swamped in the breakers and all would have been drowned.

Always cold, Winter or Summer, it is regarded as little less than miraculous how the tars succeeded in making the run up the coast in an open boat without more fatalities occurring.

SURVIVORS MAKE WONDERFUL TRIP

(By the United Press.)

Seattle, Feb. 17.—The Merchants Exchange received a wire from Neah bay this morning stating that a small lifeboat of the Emily Reed, wrecked off the coast of Oregon last Friday morning, had arrived at that place with the mate, two sailors and the body of the cook, who died early yesterday.

The men came north, a distance of over 200 miles, without a sail, and propelled almost solely by the current until off Cape Flattery, when they took up the oars and arrived at Neah bay in almost an exhausted condition.

The three survivors were taken to the lighthouse and there given medical attention. They will leave for Seattle tonight on the steamer Bellingham, arriving here tomorrow afternoon.

Ship Emily Reed – *"Another shipmate, the Reed's cook, lay dead on the floorboards, having succumbed to the terrible ravages of extreme thirst."*
Sketch courtesy of Thomas D. Turlo

"GOLD!"

John August Sutter was a short, fat, kindly man, a Swiss immigrant, who in 1834 arrived in California and set up a fort and a colony. One day in January 1848, a mechanic from New Jersey, James Marshall, while working on a sawmill for Sutter, noticed something glinting in the water. Picking up the shiny nugget, he immediately brought it to a cook who helped him boil it in lye for hours. When it failed to tarnish, Marshal gathered up some more specks and brought them to Sutter. Another test furnished the incredible truth – it was GOLD for sure. The discoverers could not keep the secret. Soon the news had reached the California coastal communities and from there it spread like wildfire east, north, and south. The ensuing rush brought people from all over the world but mainly from eastern and midwestern states where men made their plans to rush to what they hoped would be easy fortunes. 775 ships sailed for the gold fields helping to bring 40,000 new citizens to San Francisco in 1849. California's population jumped in that year from 20,000 to 100,000. And among the newcomers there were several men from Waldoboro – including a 28 year old member of Isaac Reed's family, his son Gardner K. Reed.

There were three ways for people to reach the gold fields: 1. Overland – by wagon – across plains, rivers, and mountains – this way was favored by most Midwesterners and some easterners – it was considered the most difficult and dangerous. 2. By sea to either Panama or Nicaragua, across the isthmus and then charter a boat up the Pacific coast. 3. By sea around the southern tip of South America, the fabled Cape Horn. Although t was a 15,000 mile journey and the storms as they attempted to round Cape Horn were legendary, this was considered the best way. Most easterners, including those from Waldoboro, chose a sea route.

On route number 3, the famous California Clipper Ships with their beautiful tall masts, concave bows, many sails, and relentlessly driving captains are best known in the bringing of people and goods; however, one must not forget about the hundreds of other "ordinary" vessels that were involved in this story.

We know of at least five Waldoboro built vessels that were hired for this voyage. One was the 200 ton brig *Antares* (Captain Edward House) which departed from Beverly, Massachusetts on September 22, 1849 with seven passengers. *Antares*, eleven years old at the time, had been built by James Cook. She arrived in San Francisco February 22, 1850 after a journey of 152 days.

In Chapter One we spoke of the brig *Caroline Clark*, built by Joseph Clark, which traveled with 38 passengers leaving on November 16, 1840 and arriving at San Francisco June 14, 1850. Her captain was Bradley, and she was five years old when she began her voyage. After the trip she was sold into a South American transport company and never saw the East Coast again.

A second Joseph Clark vessel, the 299 ton barque *Louisiana*, Captain A. Williams, left from Philadelphia December 23, 1848 together with 12 crew and 40 passengers and arrived in San Francisco. Like the *Caroline Clark*, she never saw Waldoboro again, ending her days in Oregon.

Reed and Welt's 164 ton schooner *Pedomonte* departed for the Golden Gate from New Orleans with 8 passengers and crew February 13, 1849. Stopping at Rio De Janeiro and Valparaiso on the long leg to Cape Horn, she arrived safely September 30, 1850. Somewhat later she was wrecked on Oregon's coast.

Edwin Achorn of Waldoboro, the only local to build a true clipper, finished his 916 ton *Wings of the Morning* in 1852. *Wings* made several trips to the Gold Fields after that time, once carrying a load of supplies to some Waldoboro prospectors. One of those prospectors was Gardner K. Reed, son of community leader Isaac Reed. In the next article we will discuss Gardner Reed, his decision to try his luck in gold, his father's efforts to diissuade him, and what luck was his in this new venture.

Sources for this article were American Heritage's *The Pioneer Spirit*; American Heritage's *Clipper Ships and Captains*; Stahl's *History of Old Broad Bay and Waldoboro*; Goodman's *Encyclopedia of the California Gold Rush Fleet.*

John Sutter
Picture from American Heritage's The Pioneer Spirit

"GOLD!" (part 2)

Mr. Gardner K. Reed (1821 – 1874) was one of the Waldoboro young hopefuls who emigrated by ship to California in search of gold. His uncle, also Gardner K., owned a vessel out of Boston which transported several other local men. In a letter of December 1849 father Isaac speaks to his brother and reminds him: "If all the men from this place (Waldoboro) go on your ship, you should give William (another of Isaac's sons) a present, say a good overcoat. He worked hard to induce them to go."

Gardner the younger's grandfather, Judge Redington, was another profiteer, shipping a multitude of supplies to the miners from New England. Isaac, in another letter January 27, 1850, describes this activity: "he (the judge) has been doing considerable in the California business with the prospect of making money. He shipped about 90,000 feet of lumber from Bath."

Meanwhile Gardner himself was trying to decide whether or not he should go at all. His father, somewhat fearful for his son's health and dubious about the potential profitability of such a journey, gently questioned him. However, early in 1850, Gardner embarked for the gold fields. His letters to his family over the next few years describe fluctuating fortunes, hopes, and disappointments. Also he spoke about the other Waldoboro men he would occasionally meet.

1857 – "I saw a man there who came to California in the same boat that I did. He told me that at one time he considered his property worth $370,000. Now it is doubtful if it would sell for enough to pay his debts. And so it is with many others."

1859 – "When I shall be able to start for home, I hardly know – if good luck should attend me, perhaps in a year or so. If bad, perhaps never. I have contended against fate for 8 long years and am neither conquered nor subdued. By "friends" and too trusting an attitude I was swindled out of my property; but I mean to accumulate property again and do it fairly and honestly. I do not fret or worry much. I ceased doing that some time ago but keep struggling on... and I think I enjoy myself about as well, living as I do in my Log Cabin, as I did when in Boston, surrounded by comfort, style, and luxury."

Gardner K. Reed never saw Waldoboro again. On January 28, 1874, while attempting to cross Rock Creek in Yuma County, he was tragically drowned. His remains were transported home and laid to rest in the Main Street Cemetery with the rest of his family.

A footnote: The originators of the Great California Gold Rush, James Marshall and John Sutter claim no happier ending to their lives than does Gardner Reed. The discovery which made some rich ruined Sutter. Neither laws nor force could restrain the prospectors who overran his land, butchered his cattle, and destroyed his buildings. By 1852 he was bankrupt. California eventually granted him a pension of $250.00 a month; but by 1878 the pension had culminated. He died two years later, still petitioning his state from a house in Pennsylvania. James Marshall fared even worse. He died in poverty selling his autograph to support himself.

Captain Nathaniel Groton's Journal

Waldoboro is famous for its giant vessels – the 1689 ton *Gov. Ames* built in 1888 by Levitt Storer and the six William Palmer built between 1900 and 1904 by George Welt being the best known. What is not so well known is that hundreds of smaller craft – schooners, brigs, barks, and barkentines were also constructed in the 16 (at least) shipyards from this village. Most of these were used for the coastal trade as far south as Latin America – carrying any cargo that would pay out of Maine and often returning with another cargo to sell. These little vessels were, in fact, the pickup trucks of the day when travel and commerce on water was carried on much more quickly and easily than travel and commerce on land.

The Maine Maritime Museum of Bath has in its archives many of the logs or journals kept by skippers of these "coasters"; among them is one compiled by Captain Nathaniel G. Groton of Waldoboro. Captain Nathaniel was the nephew of James R. Groton (pronounced "Growton") who was a village storekeeper as well as a shipbuilder on the west side of the Medomak. Nathaniel had taken command of the brig *Times* (147 tons – built by James in 1847) in May of 1849 as she coasted from New York to Waldoboro and thence from Waldoboro to Philadelphia, back to Boston, then to Savannah and back to Boston; then to Matanzas, Cuba, and finally arriving in Boston March 12, 1850.

For the readers of ships' logs, most of the writings usually concern business matters only; the weather, direction and velocity of the wind, compass readings, and anything unusual that happened. The day was divided into three parts – "First"; "Middle"; and "End". Many times the journey would be quite tedious, and the log would reflect this. However, Captain Groton, being a good writer as well as very honest about expressing his feelings, made his journal quite interesting; for example, he injected into the narrative very clearly how he was feeling – ANGER toward his uncle; HOMESICKNESS for Waldoboro; LONGING for a certain someone.

Here are some samplings from Captain Groton's journal. PLEASE NOTE: the spelling errors are the Captain's!

"Wednesday, May 2, 1849 – Commences with lite breezes from the north with fine pleasant weather. At 8 A.M. got under weigh with all sails set and proceeded down the Bay (from New York). At 11 A.M. past Sandy Hook with lite variable breezes at meridan(noon); wind east. Winds very moderate at end off Sandy Hook."

"Friday, June 22, 1849 – First part lite variable breeze with smoky weather – NNE. At 6:30 A.M. breezed up from the SW; squared yards and set steering sails; at 9 past Pemaquid Point, at meridian (noon) past through Lower Narrows; and at 3:00 P.M. brought up in the mud one mile below the town of Waldoboro. (waited there for the tide.)"

"Sunday, July 15, 1849 – Commences with fine breezes from the North with fine pleasant weather. At 6:45 A.M. got under weigh from the wharf in Waldoboro and proceeded down the (Medomak) river with a strong breeze. At 8:40 A.M. hove to off Muscongus; put father and Captain Kuhn (temporary passengers) ashore at 9:00 A.M. Took in the long boat, stowed it, and squared the yards for sea. At 10:00 A.M. past Pemquid Point with all sails set to the breeze."

"March 13, 1850 – Arrived in Boston; found my uncle J.R. Groton waiting for us with his pretty Sulena. The damned old miser – he wanted to sell the dear old *Times* upon his arrival, which he done. May he NEVER prosper – a most earnest wish of Nathaniel G. Groton!"

(Note – A most unhappy and homesick Captain Nathaniel Groton was now in command of a different vessel – the brig *Eliza Merithin* bound from Boston to Matanzas, Cuba.)

At sea — "May 24, 1850 – Latitude 30 39' North; Longitude 10 55' West – List ye, all ye heavens – Nathaniel G. Groton is far far away from the land of my birth (Waldoboro) and those that are near and dear to me; lonely pass the hours aboard this vessel. I do not feel at home as I did aboard the old *Times*; then I looked ahead, but now it is for one short voyage and then leave. But I will leave no stain behind when I depart. Short voyage did I say? The hours are months to me. Oh how glad I shall be when I am once more permitted to step my foot on the land that gave me birth and to take by the hand those that are near and dear to me. Then I think I shall be happy for a little reason; alas, how frail is man cast upon life's stormy billows, doomed to disappointments on every hand. For above me, what do I behold: the clear blue sky; beneath me the deep blue wave; and around me – a WORTHLESS crew! How can I be happy under the present circumstances and when I left the ONE who was most dear to me, she looked upon me with a frown, as if to say "I care not whether you go". And so we parted. But time does not blot out the love, nor will it ever, while life remains."

At Matanzas, Cuba — "June 16, 1850 – Discharged cargo and took on molasses. Mate very disobedient and negligent of duty. Took on board 17 casks of molasses. Preparing to leave for Boston."

Schooner Irene E. Meservey – *similar to Nathaniel Groton's* Times *in size and rig.*
Courtesy Maine Maritime Museum

The Story of the *Mabel Clark*
Part One – The Launching

Joseph Clark, the great Waldoboro shipbuilder, major employer, and wealthiest resident, died suddenly of heart failure, July 19, 1875. A large number of people were in attendance at the funeral, for which businesses were closed, work in the town generally suspended, and flags upon buildings and ships displayed at half mast. Although his death was deemed an irreparable loss to the community, the town was fortunate to have Mr. Clark's very capable son, Edwin O. Clark, to carry on the many family enterprises.

Edwin O. Clark soon launched the first of five vessels built completely under his name. This was the 1661 ton ship *Mabel Clark*. *The Lincoln County News* of 1877, as was its wont, kept and reported her progress as she was being built:

April 26: "E.O. Clark has stretched the keel of his ship, and the frames are going up."
May 10: "Mr. Clark has the frames all up."
August 9: "Captain O.D. Sheldon of Philadelphia is to command and part own the new ship built by E.O. Clark."
September 20: "The Davids Brothers of Damariscotta are preparing the rigging for E.O. Clark's new ship."
September 27: "Plankers have 'shut in' on Clark's ship which is expected to be launched next Thursday."
October 4: "Mr. Clark's vessel will be launched next Monday."
October 11: "The launch of the *Mabel Clark* took place Monday at 11 A.M. at the Clark yard. She is 201 feet by 40 by 24 and her tonnage is 1661. Her steering mechanism is of the most modern type and was built in the James Boyd Iron Foundry here in Waldoboro; her spacious 50 foot cabin displays the following woods: mahogany, black walnut, birdseye maple, California laurel, black walnut burl, mountain ash, satin wood, and California cottonwood.
A tug towed the *Mabel Clark* to Hollis Point where she will be fitted for sea. After rigging and fitting are completed, she will probably proceed to New York and take grain freight for Europe. Mr. Daniel Hitchcock of Damariscotta will ship as first Mate."
October 25: "The *Mabel Clark*, nearly ready for sea, invited a party of ladies and gentlemen on board. The company embarked from Clark's Wharf and were rowed to David Point where the vessel was anchored. The guests inspected the vessel and enjoyed a sumptuous meal at 2:00 in the spacious and elegant cabin. After another hour of quiet enjoyment, they took to the boats again and returned to the village at 6:00, loudly praising Mr. Clark, Captain Sheldon, and Mate Hitchcock for their urbanity and kindness in answering questions and affording the party every means of having a downright good time."
November 1: "The ship *Mabel Clark* left her anchorage at sunrise yesterday and was towed to sea by the steam tug *Knickerbocker* of Bath. The ship goes to Norfolk, Virginia where she loads cotton for Liverpool.
A crew of 12 men and a cook arrived in a coach from Rockland; they were furnished by William Wilson, shipping agent at 140 Commercial Street, Boston, and are said to be a superior crew."
"The ship *Mabel Clark* arrived at Norfolk last Saturday, passing through three gales safely enroute to her destination."

Part Two – The Tragedy

In the midst of a scorching heat wave in July 1878 *The Lincoln County News* published the first of a series of reports concerning the fate of the Clark ship *Mabel Clark*, launched the previous October.

"The ship *Mabel Clark*, Captain O.D. Sheldon, from Liverpool March 9 for Hong Kong was lost in latitude 37 degrees South and longitude 12 degrees West. Part of the crew was saved – no data given as yet. Later advices say she went ashore at Tristan de Acunhas, an island near the Cape of Good Hope. She was well insured and one of the best vessels ever built in these parts. The Mate, Second Mate and carpenter were from Damariscotta. We do not known who of the crew was saved."

Reports continued to come in sporadically regarding the loss of the *Mabel Clark*, only five months after her Waldoboro launching.

August 8, 1878: "There are two versions about the wreck: one says that 6 seamen were lost and that 15 besides Captain Sheldon, his wife and their child were saved. The other version says that all crew members were saved."

September 30: "A letter has arrived from the First Mate, Daniel Hitchcock of Damariscotta. In it he describes a horrifying experience:

'When we struck the reef in that night of intense darkness, it was about 2 A.M. The sea was very high, and when daylight finally came, we could see the natives on the shore unable to reach us because of the immense waves. All day we waited as the storm continued. At 2 P.M. I concluded that my chances would be better if I swam for shore. Exhausted and with a damaged shoulder, I was just barely able to make it. Captain Sheldon, William Hitchcock and 5 other seamen followed my example; some were successful, but William and another man were dashed onto the rocks and perished. Those who did struggle ashore were aided by a native who swam out with a rope tied to him, took seamen in his arms, and swam back. The rest of the crew remained on the ship which was stuck on a reef about an eighth of a mile from shore. Most climbed into the rigging to save themselves, but some were washed over the side and drowned. Mrs. Sheldon, tied to the mainmast, did not budge from her perch for 34 hours; her little 10 year old daughter had been wrapped in a rubber coat and held tightly by one of the stronger men. Once she was wrenched from his grasp for an instant, but her hair caught in one of the large blocks; he was able to lash her to a mizzen top. She survived the ordeal unharmed, except for two broken teeth.

On the third day the storm subsided; eventually all remaining crew members on the ship were rescued and brought to land. We all remained on that tiny island for the next 49 days. Then the Captain, his family and the other officers were taken to Singapore via the ship *Maulesden*. The remaining crew members stayed on the island with the 94 permanent inhabitants, grateful to be alive. Later they were transported to the Cape of Good Hope from where they will make their way home by October, it is hoped'

October 17: "Mr. Daniel Hitchcock is home in Damariscotta! He has had his shoulder set in Boston and is recuperating at his house.

Captain Sheldon has arrived in Waldoboro and confirms the Mate's account of the tragedy.

Mr. Edwin O. Clark has commenced work on a new vessel – a brig for Captain Stahl."

It was not quite a year since the happy launching of the ship *Mabel Clark*.

Ship Mabel Clark – *"On the third day the storm subsided; eventually, all remaining members of the ship's crew were rescued and brought to land."*
Sketch courtesy of Thomas D. Turlo

The Lincoln County News – May 8, 1879 – Edwin O. Clark launched Saturday morning the stanch and beautifully modeled hermaphrodite brig *Stacy Clark*. She is 393 tons and 130 X 31 X 12 feet. American Lloyds judged her A-1 for 12 years. She is owned by the builder, by parties in Boston and New York, and by her Captain, John B. Stahl. The tug *Knickerbocker* towed her Tuesday to Horse Island Harbor at the mouth of the Kennebec where ice will be taken for Fernandina, Florida, thence to Uruguay and return to Baltimore."

From Robert Applebee's records: The brig *Stacy Clark* was owned by E.O. Clark and others from 1879 – 1882. In 1897 she was owned by Peter Starkey, the Noland Bros. and others. On October 24, 1897 she foundered off Cape Hatteras.

Brig Stacy Clark *– 373 tons – built by E.O. Clark 1879 – Courtesy Maine Maritime Museum*

Mutiny Aboard the *Ambrose Light*

News from Waldoboro ships at sea would reach home weeks later because of the slow communications. (Those of us who complain about "snail mail" should take heed here!) Usually the entire story of the event at sea would arrive in bits and pieces and would have to be put together later. Such was the case of a mutiny which took place on the brig *Ambrose Light*, a 300 ton vessel built by Alfred Storer in 1857 at his yard just south of the current Town Landing.

The April 2, 1875 edition of *The Lincoln County News* gave an account of this event with the following facts set forth:
On a passage from Boston to San Juan, Puerto Rico the ship was sailing in Latitude 35 degrees North and Longitude 64 West on Christmas Day, 1874. At 8 bells both watches were engaged in setting the foresail; Captain Daniel Schwartz (of Waldoboro) was standing amidships. Suddenly a torrent of loud cursing was heard, and the steward, a black man named Francis Nickerson, was seen striking the captain from behind repeatedly with a large knife. Bleeding badly, the captain attempted to stagger aft, but his assailant followed him, brandishing his weapon, a fearsome sheathknife sharpened on both edges. By this time Second Officer Carlson had managed to grab the knife and with the help of several crew members hold and subdue Nickerson. Captain Schwartz, bleeding profusely from three gashes in his cheek and chest, was carried below and bandaged. He remained in very poor condition upon arrival of the brig at San Juan. There, a United States Consul ordered that the steward remain in irons until the vessel reached Massachusetts and a trial could take place.

A later account, this time from the Boston *Daily News*, gave the concluson to the story. (Author's note – Whether justice was truly served is left up to the reader.)

Francis Nickerson was arraigned in Boston for his attack upon Captain Daniel Schwartz. There was evidence presented to show that the steward had been inattentive to his duties in cooking meals for the crew. Just that morning the captain had announced that he had better get breakfast done in a hurry. Nickerson had then protested and had accused the captain of continually mistreating him; he added that San Juan would be the place that he wanted to and would leave the ship. With that the captain swore that he would force the steward to return to New England "dead or alive." He turned and strode back up on deck. A moment or two later came the attack.

Naval Commissioner Hallett of Boston ruled that the attack was "most unjustifiable" and ordered a fine of $3,000.00 for Steward Nickerson.

Captain Schwartz recovered from his wounds and returned home.

Brig Emily T. Sheldon – *built 1873 by Joseph Clark & Son – similar to* Ambrose Light *in size and rig.*
** Note – In the photograph she is rigged as a brigantine. Courtesy Maine Maritime Museum*

Down Easter George Curtis – *1745 tons – launched 1884 by A.R. Reed & Co.*
Courtesy Mr. Fred Storer

Ship *George Curtis* – Long Lived Down Easter

The maritime term "Down Easter" describes a type of large, sturdy, full rigged cargo carrying vessel built mainly after the Civil War in Maine. Steam and steel were inexorably winning out in the trading world; however, there was for a time, primarily in the 1870's and '80's, a need still for sail. Down Easters were fast – although not as swift as the famous clippers of the '50's – and because they were not driven so hard, were able to survive, usually, for a longer time.

This wonderful photograph of the Waldoboro Down Easter *George Curtis* is on loan from Mr. Fred Storer, himself a descendant of a distinguished shipbuilding family. It shows the *Curtis* on the ways at the A.R. Reed shipyard just prior to her launching October 31, 1877. The busy waterfront details are easily discernible as are the many buildings – most long gone – on land. It is early morning perhaps close to the 9:30 launching time.

During the building period for the *Curtis* – at 234 X 42.4 X 25.2 feet and capable of carrying 1745 tons of cargo, the largest vessel yet built on the Medomak – progress was reported weekly in *The Lincoln County News*. According to the August 22 edition "work was progressing 'finely'". The large crew, led by Master Builder Gotham C. Feyler, included men from not only Waldoboro but also Damariscotta, Nobleboro, Cushing, Friendship, Yarmouth, and Freeport. The most skilled craftsmen were earning $2.00 a day. The yard was considered one of the "most convenient" on the coast with plenty of room for all departments and with water for drinking and steam piped from a well on the Smouse property just west of the shipyard.

After the dramatic and successful launch, the *Curtis*, headed directly into the channel, was towed down river. From there she was scheduled to sail on her maiden voyage to England under the command of Captain Thomas Sprowl accompanied by wife Mary Trowbridge Sprowl. Thomas Sprowl was to remain her skipper for the next 16 years.

None of the 10 Waldoboro Down Easters ever returned to their native port. As Jasper Stahl writes: "They were deep sea sailors, globe wanderers. They sailed long voyages to strange places. Some had long lives others short and tragic ones, often leaving their bones on the shores of land far removed from the port of their birth."

The *George Curtis* was one of the long lived lucky members of the fleet, being among the last to be afloat while in active service. We have scattered reports of her after that initial trip to England – the first being her leaving Philadelphia en route to San Francisco May 13, 1887. Captain G.S. Calhoun replaced Sprowl as skipper in 1900.

From 1908 to 1910 she was a member of the famous Welch fleet (five Down Easters and two steel Scottish built barques). This fleet sailed regularly out of San Francisco carrying freight and passengers to and from the Hawaiian Islands. (The round trip price for passengers was $40.00 per person – as opposed to $75.00 on a steamer. The outward trip, frequently helped by a following wind, averaged 10 - 14 days; returning against the wind – three weeks at best.) Hawaiian sugar was a common return trip cargo.

In 1910 the *Curtis* was sold again – this time to the North Alaska Salmon Company. Many of the old square riggers were used in this trade – waiting through the winter in San Francisco harbor and then loading up with fishermen, salmon fishing gear, and canning paraphernalia. In early summer they would sail to Alaska, unload, and then wait for the fishermen to fish and their catch to be canned and loaded before returning to San Francisco. *George Curtis* survived 11 years in this business, and watched as many of the sailing vessels were gradually being replaced by faster, more reliable steam ships. In 1921 she was sold once again, for the last time – back into the Hawaiian sugar trade to work for the San Francisco firm of Libby, McNeil, and Libby.

The final news on the now barque *George Curtis* (having been re-rigged in 1913) is that in 1928 while moored in Seattle harbor, she caught fire, burned, and was abandoned. It was 44 years since that launching day in October, 1877, from A.R. Reed's shipyard in Waldoboro, Maine.

Sources of information for this article include: *The History of Old Broad Bay and Waldoboro* – JJ Stahl, *The Down Easters* – Basil Lubbock; *American Merchant Ships 1850 - 1900* – Frederick Matthews; *The Lincoln County News.*

Special thanks to David Dearborn and Cathy Matero at Maine Maritime Museum.

Special thanks also to Fred Storer.

Governor Ames *1888 – Leavitt Storer Yard, 1689 Tons*

16, 1909 ** Copyright, 1909, by Post Publishing Co. SIXTEEN PAGES—ONE CENT

14 DEAD IN WRECK OF GOVERNOR AMES

Big Boston Schooner a Total Loss---Sole Survivor Tells How Captain's Wife Was Killed by Falling Mast in Great Gale

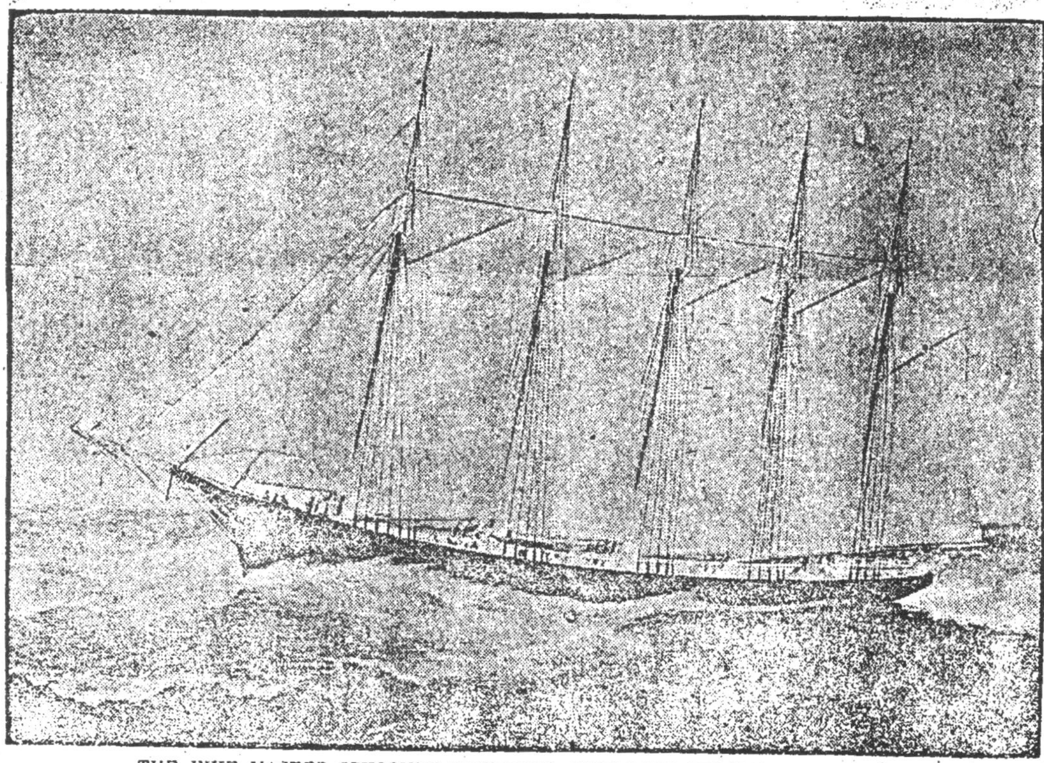

THE FIVE MASTED SCHOONER GOVERNOR AMES LOST OFF HATTERAS.

CHARLESTON, S. C., Dec. 15.—The captain, his wife and 12 members of the

For hours the crew had been working without sleep trying to keep the ship on

Speering is confident that no others could have been saved. He saw the cap-

Last Hours of the *Gov. Ames*

In an earlier entry we described the career of the famous Waldoboro schooner, *Gov. Ames*, built as the first five master in this country. Levitt Storer and his 150 man crew constructed this huge vessel – 245 feet long, 50 feet wide, 21 feet deep – at the site of the current Town Landing in 1888. After a profitable 21 year career, carrying cargoes all around North and South America, she finally met her end on Wimble Shoals, 25 miles off Cape Hatteras, North Carolina, December 13, 1909.

The following eye witness account of this tragedy was written by Josiah Spearing, the sole survivor. It was published in the *Portland Press Herald* years later by historian Mark Hennessey.

"At 10:00 on the morning of Monday, December 13, Charles Morris took the wheel. Before eight bells we were ordered to reef sail. The Captain (Nova Scotia born A.M. King) was on deck when we got there. A few minutes after starting to reef sail, the ship struck bottom.

The seas were running heavy, and the ship began to pound and break to pieces. With the rest of the crew I assisted in protecting the wife of the Captain, until the main-mast, to which she had been lashed, fell and killed her and one of the sailors, who was attempting to unlash her to give her better protection. The main-mast fell with the rest of the masts. I went onto the forecastle head with the mate. From there we went on to the jib boom, staying there until I was washed inboard. (Note – The *Ames'* jib boom was 75 feet long.) I then made my way, crawling and walking some, to the stump of one of the aftermasts.

There two or there men were hanging to the stump of the mast which was leaning aft, and one man was on top of the mast. He asked me to hold him as he said his fingers were gone. I did so until the mast went down all the way, and we were washed off.

I washed hold of some ties (Note – The *Ames* was transporting 33,000 railroad ties on this trip.) until I could catch hold of the hatch, which was floating near me. I held onto that for a while until I floated near the strongback of the main deck and ribs that had been spiked or bolted across and made a good raft. I floated on that until I was picked up by the steamship, *Shawmut*, bound for Charleston, South Carolina."

Note – The *Ames* normally carried a crew of 9 seamen plus the Captain, Mate, and cook. All, save Mr. Spearing, lost their lives.

Schooner Gov. Ames – *"There two or three men were hanging to the stump of the mast which was leaning aft, and one man clung to the top of the mast."*
Sketch courtesy of Thomas D. Turlo

·1909

EVERETT COUPLE DIED IN GOV. AMES WRECK

Capt. A. M. King of Boston Schooner and Wife Had Been Long at Sea---Crew of 12 Booked Here.

Capt. and Mrs. A. M. King, who drowned with 12 others in the wreck of the five-masted schooner Gov. Ames off Cape Hatteras Monday, made their home when ashore with their daughter, Mrs. Robert Winterbottom, 21 Russell street, Everett.

Capt. King was part owner of the store of Winterbottom & King on Broadway, Everett, and three weeks ago was at the home of his daughter for several weeks while his vessel was being loaded for her southern trip at Mystic wharf.

Capt. King was born in Sheldon, N. S. He had been at sea for 30 years and was never in a serious accident. For the past eight years he had captained the Gov. Ames, which was the first five-master ever built. The owner of the Gov. Ames was Capt. C. A. Davis of Somerset. Mrs. King passed most of her life at sea with her husband.

Ship's Varied Career.

The vessel was chartered for the trip which ended disastrously by John S. Emery & Co. of 114 State street, who are the Boston agents of the owners, the Atlantic Shipping Company of Somerset. The vessel had on board 33,000 railroad ties, consigned to G. S. Baxter of New York, and the ties were being carried from Brunswick, Ga., to New York when the vessel was lost. The schooner was valued at 30,000 and is probably insured.

The Governor Ames has had a varied career. On her very first trip she was totally dismasted in a fierce storm on the Georges fishing bank, and the owners were somewhat disheartened, but she was repaired and for many years had sailed the coast with few mishaps.

Early in 1896 the Governor Ames saved 19 people from the freight steamer Tillie, which had been converted into a Cuban filibusterer. It was charged that the Tillie had a traitor on board, who scuttled her. At any rate, she began to leak badly, and but for assistance from the Governor Ames all on board might have been lost. As it was, the steward and three Cubans went to a watery grave. The Governor Ames, by lying to, was able to take on board the 19 men who escaped. The crew of the Ames were daring on this occasion.

Names of Crew.

In May, 1899, she stranded on Low Key, near Key West, Fla., and for a time it looked as if she could not be floated and would prove a total loss. She lay on a rocky bottom. The cargo of coal was removed, she was pulled into deeper water, and after slight repairs proved as good as ever.

Her last previous serious experience was in 1903, when she was hit by a hurricane off the coast of Virginia. Her sails were stripped to ribbons and she was in a perilous condition. Fortunately, the wind shifted and she was able to make her way into Newport News.

At the office of the U. S. shipping commissioner on Commercial street, the names of the crew of the Gov. Ames appear as registered at this port on Nov. 8. The crew on that date signed articles for three months. The cook, whose name is Webb, and who was a native of Nova Scotia, aged about 45, does not appear in that list, but he was on board.

The names of the crew are on the register as follows:

Name. Birthplace.	Age.	Duty.
W. Kraft, Germany	37	first mate
A. W. Petersen, Denmark	48	second mate
C. S. Higgins, P. E. Island	48	engineer
Charles Mars, Boston	30	seaman
Andrew Anderson, Norway	32	seaman
L. Anderson, Norway	20	seaman
S. A. Johnson, Norway	52	seaman
George De Young, Africa	40	seaman
Josiah Spearing, Nova Scotia	35	seaman

From the collections of Captains Douglas K. and Linda J. Lee

Millard and Willard

Waldoboro ships were often skippered by Waldoboro men. Captains John B. Stahl, Samuel E. Stahl, Albion F. Stahl, Herman Kopperholdt, and David Hoffses were some of the renowned gentlemen who went to the sea as lowly cabin boys and advanced to command their own vessels at the age of 25 or 30. Character, knowledge of ships and the sea, and the ability to lead (abetted by quick fists) brought about this advancement. A ship's captain was expected to guide his craft to any corner of the world, regularly being away from home for one, two, and three years. When he reached his destination, he was expected to be able to barter and bargain shrewdly but honestly with merchants of different cultures and languages from his own, and to bring back products which, when sold, would make the ship owners a profit. It is a credit to these very young men that many of the vessels, by the dint of their bargaining, paid for their building costs with the profits of the initial voyage, and that later generations of the owners' families are still living from some of the profits earned in the 19th century!

Individual captains regularly signed on for a number of shares in the vessel they commanded plus a substantial salary. Their quarters on board were comfortable and spacious compared to the crew's; they were allowed to bring their families with them on some of the voyages. But the responsibility they bore for the ship, cargo, and an often unpredictable and obstreperous group of seamen was enormous – despite their seat of total power that they held on their vessel.

Two of Waldoboro's most illustrious captains were the identical twins, Millard and Willard Wade. Born about 1865, they were the sons of Captain Thomas Wade and his wife Mary Benner Wade and the brothers of another seafarer, Eugene Wade. Millard and Willard commanded at least five sailing ships each between 1890 and World War I. Two of their vessels were built in Waldoboro – the *Madalene Cooney* (Millard), and the *Ella M. Storer* (Willard). the local *Lincoln County News* often reported on their trips to Cuba, Georgia, or some other ports usually on the east coast.

Millard's last command was probably his most famous. During World War I he was the skipper of the *Felix Taussig*, a steamer transport. One night he received a warning to be on the lookout for German submarines. At two A.M. a lookout reported seeing a sub; the *Taussig* opened fire and sank her killing 17 of the 22 men aboard. She was, tragically, an AMERICAN vessel attacked in error, and following the incident, Captain Wade and three other officers were put on trial. Fortunately they were able to prove that they had been given a cease fire order by Captain Wade, and they were exonerated. This experience was enough for Millard. He joined his twin in retirement at their homes on Friendship Road in Waldoboro, and they lived out their lives with friends and family. Millard passed away in 1933 – Willard in 1936.

During their lifetimes Millard and Willard Wade enjoyed and took every advantage of their incredible likeness as twins. (It is said that even their own mother could not tell them apart and may even have mixed them up in the nursery before red and blue ribbons were affixed to the babies to positively identify them.) Numerous tales have circulated concerning the practical jokes they were fond of playing when their ships happened to be in port together. *The Lincoln County News* of August 29, 1901 included one of these. At the time, Millard was commanding the *Madalene Cooney*, Willard the

Thomas S. Dennison and both had put into Baltimore. One twin entered a chandlery's office, placed an order for supplies and then departed. A few minutes later the other twin strolled into the shop and when asked about his order vehemently denied the action. About this time the first twin appeared again to enjoy the joke with his brother and, hopefully, with the utterly confused proprietor.

Another incident reported in the article described the time when the Wades sailed together – one as Mate and one as Captain. During the voyage a seaman accused the mate of assaulting him and swore out a warrant for his arrest. The twins heard of this and plotted. At sea the captain wears shore clothing, while the mate is dressed in working clothes. When a policeman and the seaman came aboard to make the arrest, both the Captain and the Mate were dressed in their Sunday best. The seaman stood dumbfounded, as he could not tell one from the other, and no arrest was made!

(Later in life Millard Wade was to lose two of his fingers – an unfortunate event but one which made identification of the twins easier!)

Esther Soule Gross' wonderful manuscript entitled "My Father's Farm" recalls the Wade twins – "big men striding along the Village streets – or perhaps it was always the same one for they were as alike as two peas."

The Lincoln County News – July 26, 1900 – "Captain Willard Wade will have command of a new 4 master built at Thomaston by Dunn and Elliot."

Millard Fillmore Wade on left, Willard Wade on right
Photocopy of photo kindly provided by Capt. W. J. Lewis Parker

The Lincoln County News – August 24, 1899 – "The schooner *Madalene Cooney* sailed from Bath to Baltimore with Captain Millard Wade in command."

Courtesy Maine Maritime Museum

Captain Isaac Comery's Remarkable Career — Part One

Born just after the War of 1812; drawing his last breath on the eve of World War I; raising troops for and personally engaging in the Civil War; enjoying the adventures and great profits earned by Waldoboro men and her vessels; losing his fortune and his family- and then gaining them back again for his old age – this is the story of Captain Isaac Comery, consort with diplomats and merchants all over the world but always a loyal and true son of his home town.

The sea life was the great lure for young men in coastal Maine towns. Isaac, at age 13, shipped aboard one of the dozen or so packet schooners wending their way back and forth between Waldoboro and Boston. When he turned 15, he shouldered his bag and walked the 15 miles to Camden. There he joined Captain Silas Clark and his mate John Hoffses and spent the next 7 years as part of their crews on the schooners *Wanderer, Republic,* and *Lancet* as they made trips – mostly south but also to Belgium and Portugal.

At age 22 (and having impressed many) he was offered a captaincy; but feeling the need for more experience he turned down the offer and accepted a first mate's position instead. As mate he worked on the Waldoboro – built vessels – brig *Aldabaran,* ship *Massachusetts*, and schooner *Example* as well as non Waldoboro vessels – schooner *Delaware*, and *George Gifford* – mainly in the West Indian and European trades.

While on the *Example* at Au Prince, San Domingo, he contracted yellow fever; during a voyage to Cuba on *Delaware* he was driven ashore at Bermuda where the ship was repaired and the journey completed. While on *George Gifford*, Mate Comery sustained a broken leg; the vessel itself was later wrecked on her return from Cuba to New York.

At age 29 and newly married, he finally accepted the command of his first ship, the Waldoboro – built (Samuel Nash – 162 tons) brigantine, *Julia Ann*. He remained with her for two years and then took command of another Waldoboro vessel named for his new bride, *Mahala* (Howard) *Comery*. Being part owner as well as master turned out to be very profitable – in 9 months the *Mahala Comery* earned $20,000 – a princely sum in 1850!

The year 1854 saw Captain Comery take charge of another brand new Waldoboro vessel, the full rigged ship *Alfred Storer* (Storer and Hovey – 1200 tons). For seven years he commanded this very large vessel – mostly for the cotton trade – southern ports (particularly Mobile, Alabama) to Liverpool, England. On her first voyage, the ship was boarded by a heavy sea and thrown upon her beam ends, taking in seven feet of water. Captain Comery risked his life to make sure that leaks were plugged, the water pumped out, and the vessel saved. On another voyage the ship was hit by a hurricane near the Azores. One enormous wave carried the captain and a lifeboat overboard; another wave very luckily deposited them unharmed back on deck! Arriving in Liverpool once, the ship collided with the *Alice Counce* of Thomaston and was damaged badly; Captain Comery proved himself extremely adroit dealing with not only repairmen but with legal representatives – once including those from the all-powerful East India Tea Company who attempted unsuccessfully to intimidate him.

Captain Isaac Comery was, in fact, in the 1850's, at the top of his profession – a rcspected, welcomed presence in the North, the South, and in the ports of Europe. He was comfortable dining and conversing with senators, governors, ministers and diplomats; he was also making a very tidy fortune for himself and for his employers be they ship owners or cotton merchants. However, all of this was to change with the first shots fired on Fort Sumter, Charleston, South Carolina, April 12, 1861.

Captain Isaac Comery – Courtesy Waldoborough Historical Society

Captain Isaac Comery's Remarkable Career — Part Two

Throughout the 1850's Captain Isaac Comery's main duties in his full rigged ship the *Alfred Storer* were to bring Southern cotton to English mills. He was a welcome and trusted figure not only to American plantation owners and politicians but also to British and European ministers, diplomats, mill owners and merchants. His personal fortune was growing quickly; all seemed well. However, enormous changes were about to occur.

In the fall of 1860 the *Alfred Storer* sailed with a cargo of cotton, as usual, from Mobile, Alabama to Liverpool, England. She would be the last Northern vessel to sail this route for four long years. Upon his return, Captain Comery learned of the election of Abraham Lincoln and shortly thereafter of the secession by 7 and eventually 11 Southern states in protest. His friends the cotton dealers confidently predicted that the North would never fight to correct secession; they even offered to buy Comery a brand new ship to continue their business. How surprised they were to hear his answer to their prediction and proposal: Comery turned down their offer declaring that the North WOULD fight; furthermore, he said he was returning home to Waldoboro to enthusiastically support his government. He made a trip to Liverpool and then back to Maine where the *Storer* was sold.

During the subsequent Civil War, Comery, after refusing an offered commission in the U.S. Navy, worked for the Northern cause in his home town. In 1862 he exerted all of his influence toward promoting enlistments in Waldoboro by prevailing upon General O.O. Howard, recovering from a severe battle wound, to come and make a recruiting address. These efforts resulted in 50 men forming part of Company E of the 20th Maine. A few weeks later a new company, Company A of the 21st Maine, a 9 month enlistment, was mustered in at Augusta – its captain – Isaac Comery. The 21st Maine soon shipped out to Port Hudson, La., where the men, racked with disease as much as by Confederate bullets, nevertheless played a large role in the opening of the Mississippi River by the Northern army and navy. Although sick himself, Captain Comery was there in active support of his comrades.

After mustering out following the 9 month enlistment, Isaac Comery, citing health reasons, turned down an offered post as the commander of a Union battery. He did agree to serve one year as captain of a revenue cutter in the Waldoboro district. When the war finally ended, however, it was back to his old love, deepwater sailing, as master and part owner of the 348 ton brig *Mary C. Comery*, built by William F. Storer, Alfred's brother. In the brig he made voyages to the Mediterranean and the West Indies, familiar places to Isaac. Unfortunately for him the times for wooden sailing ships and his remarkable good luck were changing.

An economic recession, a personal bank failure, and unsatisfactory shipping returns seem to have deprived Captain Comery at this time of much of the fortune he had accumulated before the War. He sold his Waldoboro estate and went to live on Harbor Island in Friendship from 1868 to 1874. At the end of this period, however, things had improved. The U.S. government had awarded him a pension for his good deeds during the war, and he was able to move back to Waldoboro. He also had taken himself a new wife – Melvina (Castner), age 23. She was to present him with a daughter, Jennie, born in 1886 when Isaac was 67!

One more vessel was his to command in 1874 before retiring for good. This was the little 87 ton scow/ schooner *Achorn*, the last craft constructed by Waldoboro's sole clipper ship builder, Edwin Achorn.

After moving back to Waldoboro with his new wife, Isaac Comery settled into retirement. He joined the newly formed Charles Keizer G.A.R. post for Civil War veterans as well as the American Society of Shipmasters. He lived on, a respected figure in town, until he was 95 years of age, finally passing on in 1914. He now lies at rest in the German cemetery with his children and second wife.

(Information for this article came from Miller's and Stahl's histories of Waldoboro and from the *Historic Record and Biographic Roster of the 21st Maine Volunteers* written by Adjutant Joseph T. Woodward.)

The Lincoln County News – December 5, 1901 – "The tug *Knickerbocker*, top photo, towed the five masts for the Palmer-Welt schooner up from Bath and through the ice of the Medomak. The tow boat *Seguin*, lower photo, is also at Waldoboro, having plowed her way successfully through the 10 inches of ice."

These two Bath built vessels were employed for many years up and down the Maine coast fetching supplies for the shipbuilders and towing vessels up and down the rivers.

Photos from Tidewater Ice on the Kennebec River *by Jennie G. Everson*

Chapter 3
Living in the Village

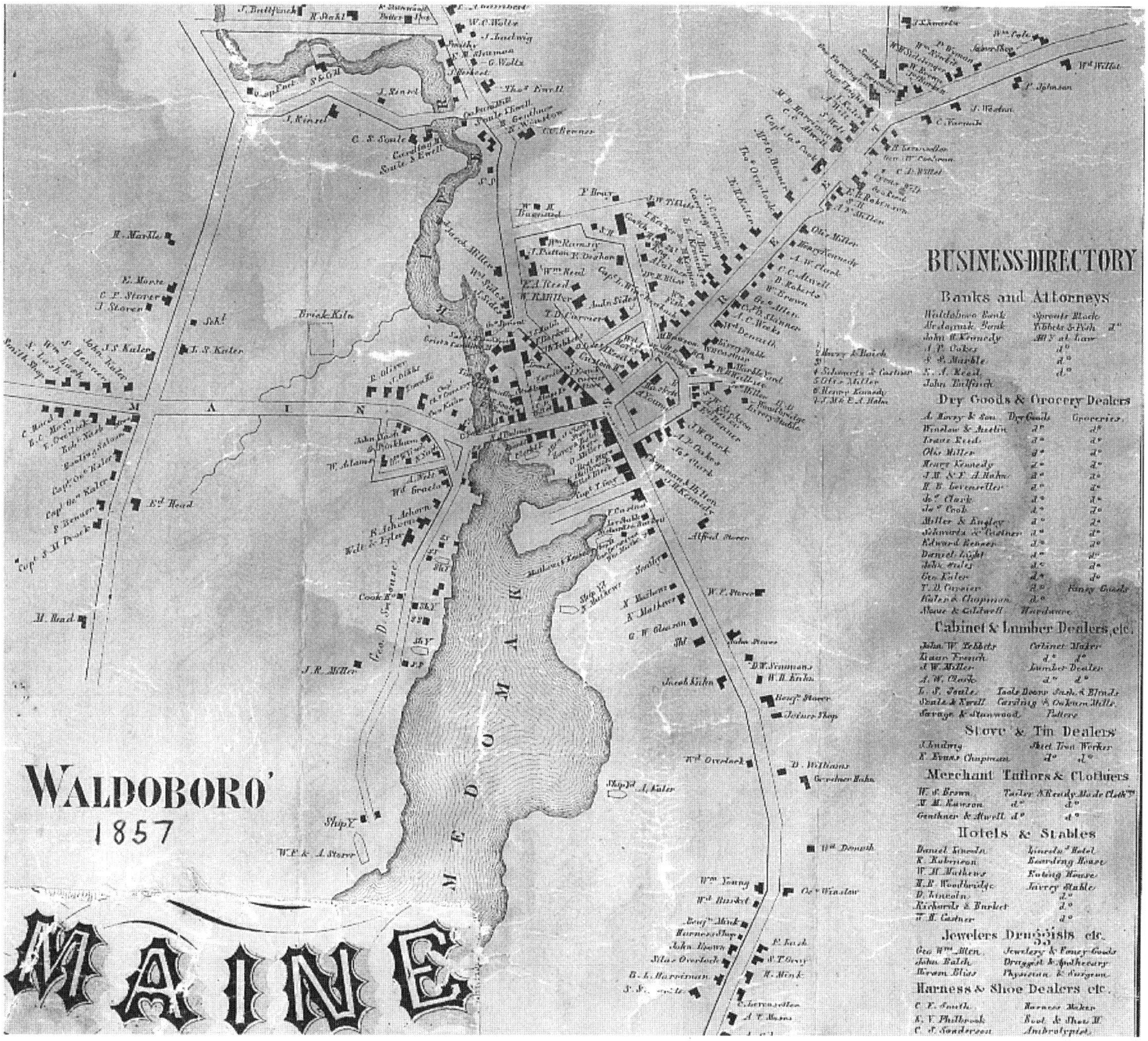

Waldoboro 1857
Courtesy Mr. Steve Cartwright

Key to the Map of Waldoboro Village

1. Sproul's Mill
2. Main Street Bridge
3. Jones Sail Loft
4. Clark's Hall
5. Fish Block (Union Block)
6. Medomak Hotel
7. Sproul Block
8. Customs House
9. Clark Shipyard
10. Hotel Savoy (Lincoln, Waldoboro)
11. Hotel Davis
12. M.M. Richards Co.
13. Baptist Church
14. Four Business Blocks
15. "Four Corners"
16. Balch's Corner
17. G.A.R. Hall
18. Town House

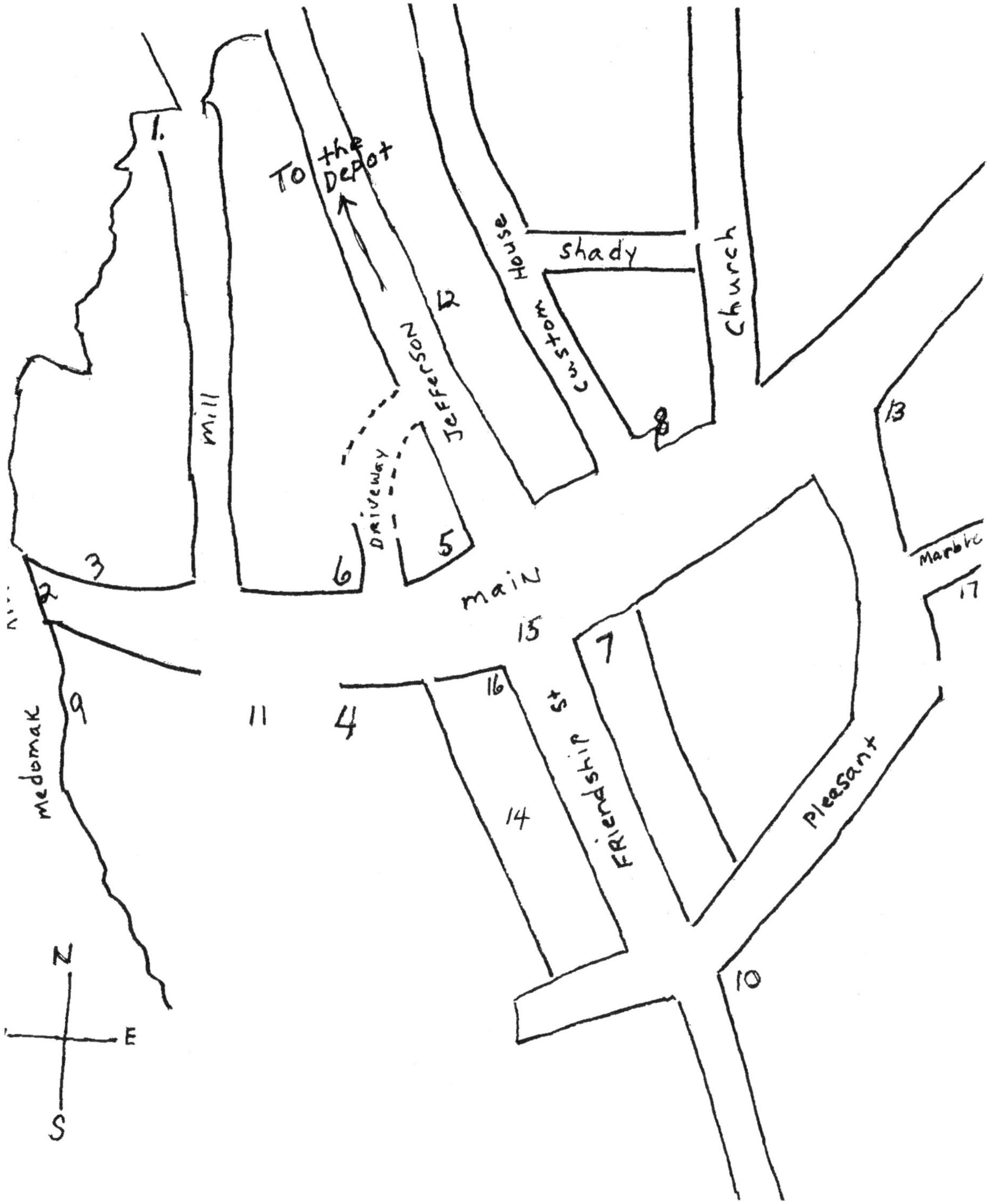

Waldoboro Village – Structures mentioned in this chapter

Friendship Street business blocks built from the ruins of the Great Fire of 1854
Courtesy Waldoborough Historical Society

The Business Center's Rebirth – 1854 - 1855

On the afternoon of August 25, 1854, a catastrophic fire destroyed many of the wooden buildings making up Waldoboro Village. Quickly the conflagration spread over to both sides of Friendship Road east as far as where the Brick Schoolhouse is today; to the westward it stopped only at the river, having engulfed one bark and one ship on the stocks as well as the lumber for another large vessel. In less than 40 minutes from its discovery many portions of the town were enveloped in one sheet of flame, and the entire eastern section was soon to be a heap of smouldering ruin. At a meeting the following evening a group of citizens estimated the amount of property damage to be in the vicinity of a half a million dollars with only a third covered by insurance. 70 families were judged homeless.

The days that followed brought onlookers to Waldoboro – to gawk at the destruction – but also to bring aid in the form of food and clothing. Meanwhile the townsfolk were tending to their own needs, and plans were already afoot for the rebuilding of the Village. Thanks to the shipbuilding fortunes that had been and were being made, there was capital available, and, incredibly, by the summer of 1855 – the new Waldoboro – this time more brick than wood – had risen from the ashes.

On the west side of Friendship Road were soon constructed the four brick blocks currently standing – financed by businessmen Thomas Gay, John Willett, E.V. Philbrook, William Storer, Francis and Edward Hahn, Henry Kennedy, Otis Miller, James Schwartz, Daniel Castner, and Charles Sampson. On the east side of Friendship Road, George Sproul built a single, long block which ran from the four corners to Dog Lane (Pleasant Street).

Through the influence of Isaac Reed the younger, at the time a representative in the 32nd Congress, an appropriation was secured to purchase a site for a Customs House (later a Post Office and much later, a Library). Mr. Reed donated a portion of his own land – an apple orchard – and later in 1855 the new building was completed.

The south side of Main Street all the way to the river was owned by shipbuilder Joseph Clark. Here Mr. Clark built a large dwelling for his daughter Celeste, a wooden business block, and finally, close to the four corners, a building known for many years as Clark's Hall – used for professional and amateur dramatics, musical shows, lectures, and travelling performers.

Across Main Street a new wooden, 30 room hotel, the Medomak House, was put up, and on the corner of Main and Jefferson Street appeared the "finest building in Lincoln County" – the Fish Block – granite and brick, four floors with stores and shops, and another hotel.

On the river there were several enterprises unhurt by the great fire. Next to the bridge on the east side stood Stephen Jones' sail loft. Along the east bank above the bridge at the first falls Sproul's Lumber and Grist Mill had been a fixture for 75 years; close by and using those same falls for a power source was the James Boyd Iron and Brass Foundry which produced farming and ship parts. At the second falls up river Charles S. Soule owned a carding mill on the west bank and an oakum mill on the east bank. At the third or "Great Falls" the Kinsell Flour Mill was standing.

And so, by 1855, the Villagc had bccn rcborn – somewhat in familiar wood, but much in more permanent, lasting brick. Great credit was due to the businessmen who, rather than pulling up stakes and abandoning Waldoboro for other places, chose to remain, build, hire, and continue. They and the others who came were soon to bring their town to the zenith of its prosperity.

Origin of the Customs House

American independence declared in 1789, the new First Congress immediately set about enacting laws designed to command attention from the old nations while increasing revenue for the new one. On the seas taxes or duties on foreign ships using American ports were stiffened while eased for American vessels. For instance, an American built vessel owned by an American firm entering an American port paid a tonnage tax of six cents per ton; while an American built vessel but owned abroad paid thirty cents per ton, and a foreign built and owned vessel paid FIFTY cents a ton. To make certain these fees were paid, a system of customs districts was set up along the coast and collection officers were appointed. Maine had 13 such districts: York, Saco, Wiscasset, Penobscot, Machias, Kennebunk, Frenchman Bay, Bangor, Belfast, Passamaquoddy, Portland, Waldoboro, and Bath. When a ship – foreign or domestic – entered Maine waters heading for a port to unload its cargo, the captain would sail either to the official port of entry and register there, or, if that were inconvenient, he could sail to a nearby harbor of a "Deputy Collector" where his business would be transacted with the deputy responsible for transferring money and paper work to the central customs house – such as Waldoboro.

In 1789 and for a few years following, there were two areas of Waldoboro vying for the honor of becoming the first Customs House site. One of these was the area of Slaigo Brook and Thomas Hill where a settlement had been started on Sampson's Cove. Here there were lumber and grist mills, a tannery, storehouse, homes, and a ferry service for the bay. Ships regularly arrived to take away wood, hides, grain, and fish and bring in imported goods from Europe by way of Boston. Squire Waterman Thomas was the major land owner and controller of commerce. It was no surprise to anyone in Waldoboro when Squire Thomas was appointed the first "Inspector of Revenue" and where the first Customs House was built – a small wooden building – on the east side of Slaigo Brook and at the foot of Thomas Hill. Thomas' appointment and the official location did not last long however; deficiencies were found in both place and man. Only intermittent seasonal water power was available for the mills from the brook and extremely shallow water in the cove for the vessels. And Squire Thomas had encountered severe financial difficulties which hurt his standing with the U.S. Government. By 1799 after a term of four years he was removed, and the Customs House site was transferred to the up and coming Head of the Tide and Waldoboro village with its almost continual water power and deeper tides. Joshua Head was appointed the Inspector. The new House was a wooden building on the north side of Main Street. Shipbuilding and commerce flourished in Waldoboro's customs district extending from Bristol to Thomaston. By 1850 tonnage produced in this area exceeded all other districts excepting Bath's.

Mid century fires – 1846, 1849, and finally – and catastrophically – 1854 – destroyed the wooden buildings of the Village – including the Customs House. Shipbuilders such as Joseph Clark, Alfred Storer, Isaac Reed, and Henry Kennedy, enriched by their trade, might, understandably, have abandoned Waldoboro and gone elsewhere. Instead, they chose to remain and help pour money into the quick rebuilding of the town – primarily in brick rather than wood. In 1855 the Federal government purchased a lot of land – an apple orchard on Main Street – from the Honorable Isaac Gardner Reed on which a new brick and granite Customs House/Post Office was erected. In 1908 the building was enlarged and very much improved. From the beginning it was built like a bank – rugged and impenetrable; payment from the captains was most often made in gold bullion.

By 1900 wooden shipbuilding, Waldoboro and Maine's lifeblood – and the raison d'etre for the Customs House system – was exhausted. The Customs Houses with their inspector were hardly needed. Locally, a movement started from Rockland in an effort to shift the official site to that busier town. Behind the scenes, Waldoboro fought hard to escape this disgrace – and succeeded. But in April of 1904, the U.S. Secretary of the Treasury announced a list of 45 districts which were no longer self supporting. Waldoboro was on that list.

With so little traffic on the Medomak River, the business of collecting customs soon was a thing of the past. As a post office, the old building continued through the 1920's, '30's, '40's, and '50's. But when funds were approved to build a brand new Federal Post Office, there seemed to be no future for the now 100 year old Customs House.

Sources for this article include *the Maritime History of Bath, Maine and The Kennebec River Region* by Baker, *Coastal Maine* by Duncan; Stahl's and Miller's Histories of Waldoboro; *The Lincoln County News*; and conversations with Delia Mohlie and Nathan Lipfert, librarians at The Waldoboro Public Library and the Maine Maritime Museum.

Main Street Waldoboro – date unknown
Courtesy Waldoborough Historical Society

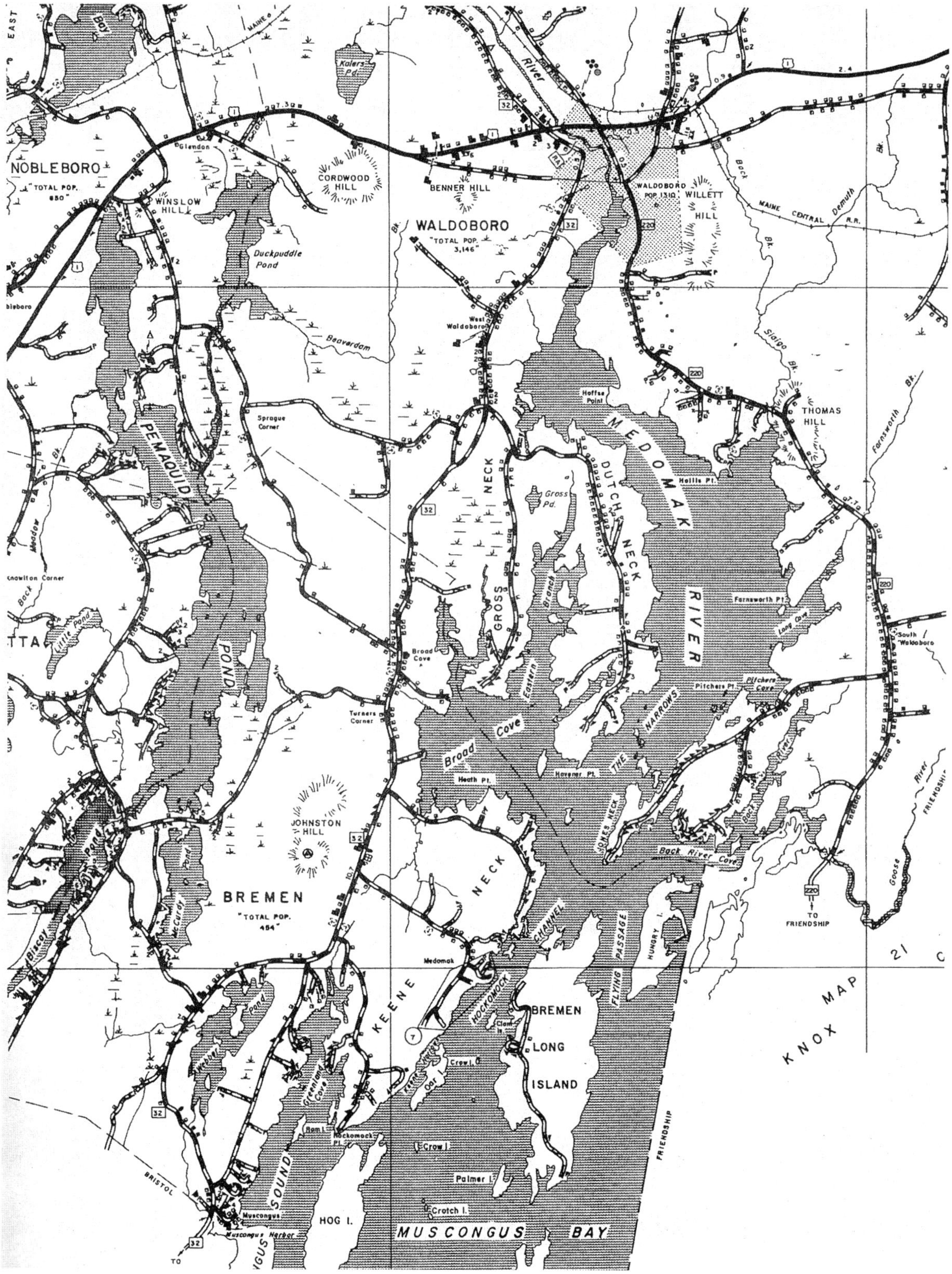

Map courtesy of the Waldoboro Town Office

A Labor Day crowd gathers at the Waldoboro Customs House/Post Office September 1906. A. R. Reed's Mansion stands to the right. Thirty years later the Mansion would be torn down to make way for the Waldo Theatre.
Photograph from the Waldoboro 1973 Bicentennial Book

Customs House / Post Office

The number of Maine Customs Districts had expanded from two in 1789 (Portland and Bath) to thirteen by 1840 (Portland, Bath, Bangor, Wiscasset, York, Penobscot, Machias, Kennebunk, Saco, Frenchman's Bay, Belfast, Passamaquoddy, and Waldoboro). These "Ports of Entry" each was provided with a government building from which politically appointed directors worked, responsible for keeping track of imports and exports, collecting duties, measuring newly constructed vessels, licensing, keeping lists of seamen, and much more – including, with the assistance of the U.S. Coast Guard – intercepting smugglers! Each district also had "Deputy collectors" at nearby places to assist in these huge jobs; Waldoboro's deputy ports, or "Ports of Delivery" were located at Nobleboro (now Damariscotta), Bristol, Warren, Thomaston, Cushing, Camden, and Rockland. The entire district comprised "all the shores and waters from the middle of the Damariscotta River to the south side of Northport on Penobscot Bay." By 1853 this district was handling more vessels than any other district in New England excepting Portland and Boston!

The beautiful federally financed Customs Houses built in Wiscasset and Waldoboro still exist today, of course. Deputy customs houses, not built by the government, were located in homes or store fronts and were not preserved.

Nobleboro historian George Dow writes that by 1888 the amount of money collected at Waldoboro had fallen off considerably as the wooden ship industry declined. Only $54.00 were paid by 24 schooners that year. By 1913 this figure had fallen even more to $47.00 from 13 vessels. These low figures in 1913 here and elsewhere justified federal action taken to close all customs houses in Maine and consolidate all business in Portland.

Waldoboro's lovely Customs House, built out of the ashes of a great fire in 1855, remained open as a Post Office after 1913. Many people in town today remember it well – as a very busy – and often times noisy – place. One entered it through the front door only; an expansive lobby with a single barred window directly in front of the door stretched the width of the building. In the lobby were rows of narrow letter boxes which were fed from the rear. Mail would be delivered twice a day (Stan Hanna remembers that Jesse Benner's team of horses would transport it from the railroad station to the back door). Upstairs (inaccessible to the public) were federal offices including one occupied in the 1950's by ex Governor – then Senator Frederick Payne who lived (conveniently) right across Main Street. Stan remembers a Lawyer Mason's office upstairs also.

Behind the mail boxes, and, except for the person in the single window, unseen by anyone, worked the postal employees. Chuck Begley speaks of a secret slot in the second floor wall through which a Post Office official occasionally used to spy on the workers below.

As was said, the lobby of the Post Office could become noisy and sometimes a little rowdy – particularly at evening mail delivery times. Mr. Samuel Miller, Editor of the locally produced *The Lincoln County News* once printed a letter from an irate citizen on this subject:

December 17, 1875: "*There has been much complaint of late that the rowdyism exhibited in the lobby of the Post Office in this village upon the arrival of the 6 o'clock mail is an evil which ought to be stopped. Boys and girls fill the place and keep up a continuous uproar until the mail has been distributed. Is this a good place for girls to improve their morals? Parents, if you send your daughters to the office, wait a half hour and avoid the bad associations to be met during mail distribution. It would seem as though the Post Master, or somebody, ought to be required to enforce better order during the short time that the crowd occupies the lobby.*"

The author respectfully thanks the following for their help with this article: Charles Begley, Allan Benner, George Dow, and Stan Hanna.

Elm (Bulfinch, Lovell) Street – Author's Collection

Squire John Bulfinch

Bulfinch, Lovell, or Elm Street

The little dead end street off Jefferson and paralleling the north bank of the Medomak River is quiet today – very quiet. Few automobiles rumble on it; there are only three homes; despite its name – ELM Street – most of the trees are small and rather insignificant. It is an EXCELLENT place to walk – peaceful and silent.

100 years and more ago Bulfinch or Lovell Street, as it was called then, was hardly peaceful and silent. As one entered from Jefferson Street, three thriving businesses could be observed by the traveler. In winter time Elden Welt's crew of ice harvesters could be seen working on the river, cutting the great blocks into two foot cubes, and storing them packed in sawdust in a nearby ice house. On the same left side of the street stood the blacksmith shop of Ephraim White, probably the most skilled of the many Waldoboro smithies. White was noted for the fact that he had designed a fine pistol when he was only 11 years old. On the right side of the street entrance James Stanwood, who had fought with the 20th Maine at Gettysburg, ran his pottery shop, often using the clay gathered on the Isaac Kaler property north east of the Corner.

Progressing further up the road the traveler could look across the narrow river at the Upper Falls and behold the Medomak Flour Mill while listening to the mighty grinding wheel turn in the rushing water. A little further he would come upon one lovely home – and then another – the houses belonging to two of the town's prominent citizens – Squire John Bulfinch and his son-in-law Captain Harvey Lovell. John Bulfinch (1791-1884) was a Bostonian – Harvard College 1812 – who had arrived in Waldoboro in 1823 after teaching school and reading law in the Maine towns of Belfast, Warren, and Union. A man of small stature, he became a giant in the fields of education and law. He was an advisor to General Henry Knox, an eminent attorney, the founder of a private academy, and a respected scholar in the classics. Investing in land, houses, ships and businesses, he supplemented his family fortunes and lived as a true "Squire" in his home with his wife Sophronia (Pike – of Camden) whom he had married in 1825.

John and Sophronia's daughter, also named Sophronia, had fallen in love with and married, in 1851, a 34 year old dashing clipper ship captain, Harvey Lovell. Lovell, from Barnstable, Massachusetts, had gone to sea at 13, become an officer at 19 and a captain shortly thereafter. In his career, he had commanded many vessels, including the speedy Waldoboro clipper *Wings of the Morning* of Edwin Achorn. Becoming a part owner of *Wings* he had come to Waldoboro and had met another part owner, Squire Bulfinch and his daughter. Shortly after their wedding, Captain and Mrs. Lovell completed a beautiful home across the street from her parents. Lovell continued his career on the sea until 1873 when he retired. Squire Bulfinch passed away in 1884; Lovell in 1899.

At the end of Elm Street – often on maps or post cards called "Bulfinch" or "Lovell" street – a bridge crossed the Medomak. (Its granite pilings can easily be seen today from the little picnic grounds on Bremen Road.) In 1907 this bridge was changed from wood to iron. Thirty years later, during a huge flood, the bridge went out.

The opening of the new Route One in 1935 and the NEW bridge there across the Medomak River had made the Elm Street bridge expendable, and it never was replaced. The pottery, blacksmithing, ice harvesting, and grist milling businesses were also items outdated by modern conveniences. For the residents, everlasting peace and quiet for their little road had arrived.

Sources for this article include: Stahl's and Miller's Histories of Waldoboro, The 1973 Waldoboro Bicentennial Book, and a conversation with Mr. Richard Wallace.

Lovell's Bridge – Author's Collection

Street scenes – Author's Collection

Town House being moved – 1938 – Waldoboro 1973 Bicentennial Book

Forgotten Buildings (Part 1)
Waldoboro's Town House

In the early years of 19th century Waldoboro the immediate West Side of the river was as much a place of business as was the East. Here, also, were shipyards, wharves, warehouses, and stores. Here two major land owning families dominated: the Head brothers, John and Joshua, who held much of the property north of the present West Main Street from the river to Kaler Pond, and George D. Smouse who owned much of the land south of the street and a busy wharf on the river. (Smouse's widow was to marry another land baron, Isaac Reed the Elder in years to come.) Joshua Head eventually built the mansion at Kaler's Corner lately known as the Snow Turtle Inn. The two families also operated stores across the street from one another at the crest of the first hill and at about the junction of the present Medomak Terrace. Both families were heavily involved with ships and shipping.

The first post office was located at the Head's store, and John Head was appointed first postmaster in 1795. Next door to the post office was the courthouse which the town purchased and moved there. In 1803 this structure was made over into a Town House – a place where records were kept, citizens could meet, and local trials could be held. This Town House stood on its spot for over 130 years and served as a voting place, a customs house (until 1855), and a site for formal town meetings.

The wooden building was small – the size and shape of a medium sized barn, remembers Martha Boggs, with benches around the interior, remembers Richard Wallace. When it proved to be too tiny to house the expanding population, it was closed, but not until the 1930's. In 1938 the town voted to sell the building to Mr. Merton Winchenbach for $100.00; it was moved a few hundred feet west and transformed into a Chevrolet garage with one bay and one mechanic, remembers Stan Hanna. In 1948 a chance spark from an electric welder ignited some gasoline and oil, and the Town House burned to the ground.

Sadly, not only is this historic building gone, but also are most of the town records once housed there and dating back to 1793. When the building was sold and was about to be moved, trucks full of irreplaceable and priceless documents, many written in longhand, were driven to the river bank where the papers were unceremoniously dumped. A scant few were rescued for future generations.

(Martha Boggs' father, along with Mr. Mank and Mr. Burns, long time selectmen of Waldoboro, moved their offices over to a tiny building on Glidden Street and carried on in the pre-town manager years.)

Information for this article was gathered from J.J. Stahl's and S.L. Miller's Histories of Waldoboro, *The Lincoln County News*, and from conversations with Ms. Boggs, Mr. Hanna, and Mr. Wallace.

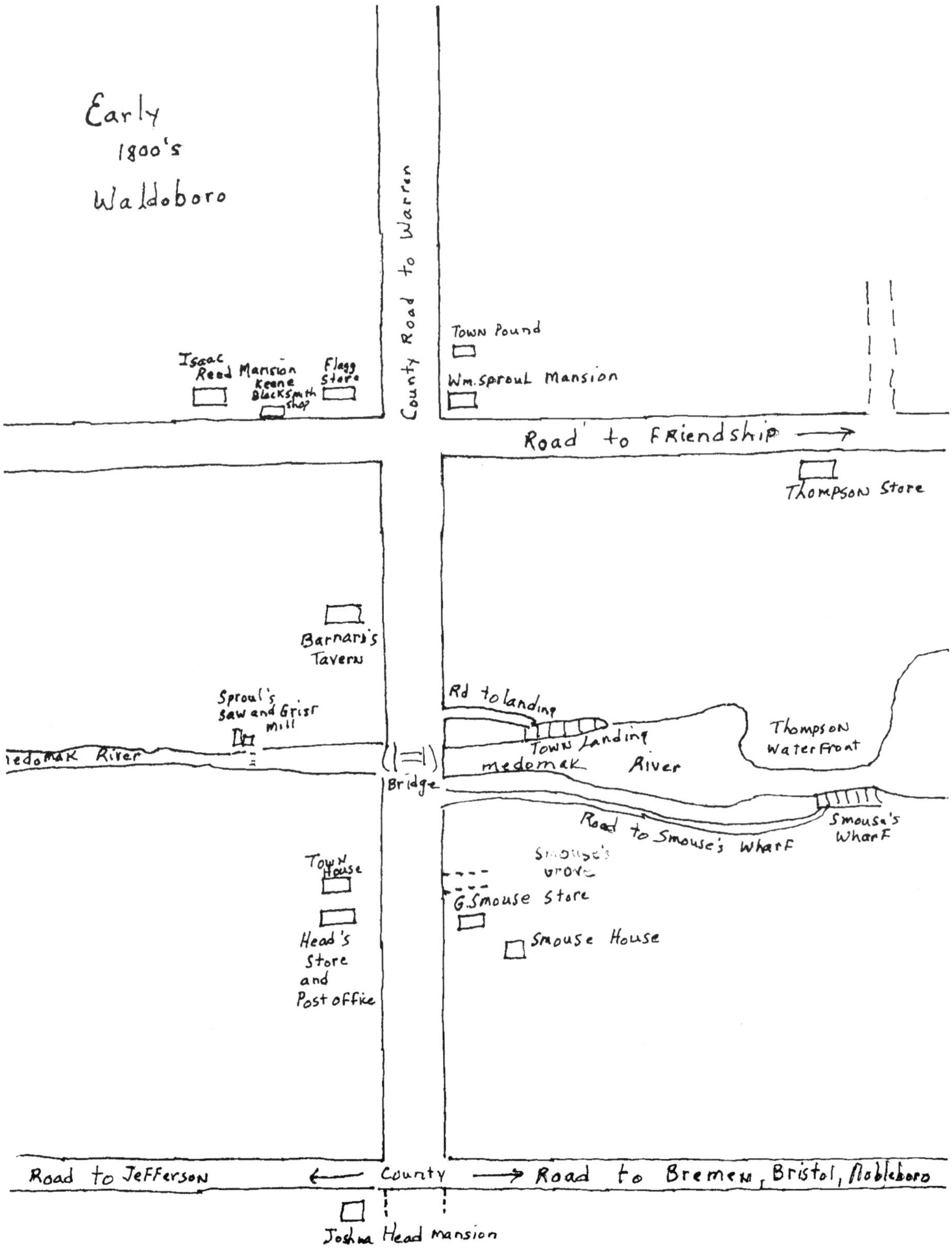

From a map drawn by J.J. Stahl

The Reunion of the 20th Maine

Many of us know that soldiers from Maine made up an extraordinarily large part of the Northern army during the Civil War. But it is not so well known that bluecoats from Waldoboro played a particularly important role one of the war's key battles – Gettysburg, Pennsylvania July 1, 2, 3, 1863.

The 20th Maine – a regiment made up of several companies including two from Lincoln County – has become probably the most famous of all Civil War regiments because of its successful defense of the wooded hill called Little Round Top on July 2. On that day Colonel Joshua Chamberlain and his men were able to withstand with muskets and bayonets several desperate Southern charges – charges that if they had been successful, would have given the Rebels an opportunity to place cannons on the heights and fire unimpeded on to the entire Northern line. There is a good chance that, had this happened, the Union line would have broken, the battle lost, and perhaps even the War itself. Chamberlain, who had been a Bowdoin College professor prior to the War, was soon promoted to General and still later became Governor of Maine. His fame, enhanced by writings about the battle, has become even greater in the last several years because of the stirring motion picture "Gettysburg" produced by Ted Turner.

For the people of Waldoboro, the 20th Maine and the story of Little Round Top are particularly meaningful: at least 30 members of the community took part in the battle. The two Lincoln County companies were Company E and Company G; most of the Waldoboro boys joined Company E.
To form an army the U.S. government issued a "levy" on all the states to raise a certain number of volunteers. In Waldoboro the public response was enthusiastic, and at a town meeting it was voted to offer $100.00 to each enlistee and to appoint three persons Orderly Sergeants to encourage the recruitment process. As a further inducement a group of citizens raised $650.00 to pay $10.00 to each enrolled person.

By August of 1862, 50 volunteers were in the fold. These would unite with 30 men from Bristol and 20 from Union to form Company E of the 20th Maine Infantry. The 20th Maine was mustered into U.S. service on August 29, 1862, fought in a total of 19 battles Antietam to Appomattox, and was mustered out in early summer, 1865.

On June 17, 1865, 11 survivors of the Waldoboro quota, after three years service, arrived home and were provided with a reception in the Union Hall (Fish Block). Five others had been killed, two had died in Confederate prisons, two had deserted, and the remainder had previously been discharged from wounds or disease. Private Charles Mero, one of the distinguished and honored 11 later remembered:

"It seems but a day ago that gallant fifty, or Waldoboro's quota for 'Three hundred thousand more', left Waldoboro with their very souls fired with patriotism, amid cheers and tears, with flags flying; and then those years when you and I, and all of Company E that are alive remember so well. And then that procession of eleven 'relics' marching down Sproul's Hill in June, 1865 to the tune of Yankee Doodle and the reception we received. All these events seem but a dream at the present time."

In Aiugust of 1896 – 33 years after Gettysburg – Mr. Samuel Miller, editor of *The Lincoln County News*, but previously an Acting Sergeant Major in Company E, arranged for a grand reunion of the entire 20th Maine contingent – much more than little Company E – at Waldoboro. Announcements appeared in state newspapers for several months prior to the event which was scheduled for the 21st. Governor Chamberlain agreed to attend along with many officers and enlisted men. (Mr. Miller made it PERFECTLY clear that this was to be a meeting for survivors of the 20th Maine only – and NOT for any other Union Army veterans!)

The following are excerpts from the August 27, 1896 *Lincoln County News* (Mr. Miller's paper) reporting about the reunion:

"Invasion of Waldoboro!!! The Town Capitulates Without Firing a Gun!!!

Friday morning a body of veteran troops appeared on the outskirts of Waldoboro village and after some preliminary arrangements, marched into town and held complete control during the day and into the evening. On the arrival of the 10:06 train, bringing Gen. Joshua Chamberlain, Gen. John Marshall Brown, and many other comrades and ladies, a regimental line was formed at the station by the Association officers who were mounted. The old battle flag, was carried by brave old Sergeant Tozier who bore it through the terrible struggle at Little Round Top. The colors were guarded by two files of men with Enfield rifles, and on either side marched Sergeant Wyman and Corporal DeWitt.

Escorted by a detail from Waldoboro's Charles Keizer G.A.R. Post, and the Town Cornet Band, the column marched up Depot Street, Jefferson Street, and across Main to Friendship Road to the residence of Samuel L. Miller, where the veterans enjoyed themselves socially until noon lunch.

At 2 P.M. the annual meeting of the 20th Maine Regiment Association was held in the G.A.R. Hall on Marble Avenue, where 100 members responded to roll call. At this meeting nominations for a board of officers were accepted and the men elected; reports were heard from various committees; letters from comrades unable to be present were read and a list of recently deceased members heard. Finally, plans for the next reunion were announced. Just prior to adjournment the names of all present were given to Secretary, Samuel Miller. Representatives from Companies A, B, C, D, E, F, G, H, I, and K were present with Company I leading with 29% present. Of Company E from Waldoboro, 9 out of the original 30 said "Here!" to the muster roll.

During the afternoon a band concert was presented from a bandstand across the street from the Customs House (Library). Carriages were provided for the lady visitors to ride around the town.

Major General Daniel Butterfield arrived on the 4:32 train from New York in time to witness, along with a very large crowd, the "Dress Parade" on Miller's lawn at 5:30. Gen. Chamberlain assumed command of the parade, with Gen. Brown as Adjutant, Major Melcher as Sergeant Major, and Lieutenant Miller as Drum Major. The Drum Corps was composed of local talent and did remarkably well."

(Author's note – The inclusion of Generals Butterfield, Chamberlain, and Brown, plus two others – Ellis Spear and William LeDuc at this reunion made this, in all probability, the most distinguished group ever to visit Waldoboro at the same time. Gen. Butterfield commanded the 5th Corps at the Battle of

Fredericksburg and was Chief of Staff of the Army during he Chancellorsville and Gettysburg campaigns. Gen. Chamberlain was terribly wounded while leading a brigade in front of Petersburg and was promoted to Brig. Gen. on the spot by Gen. Grant. He commanded the troops who received the surrender of Lee and was four times elected Governor of Maine. Gen. Spear, besides his war record, held the position of Commissioner of Patents at Washington. Gen. Brown was the first Adjutant of the 20th, promoted to Assistant Adjutant General, and Colonel of the 32nd Maine. General LeDuc was a famous Quartermaster General during the War and was Commissioner of Agriculture under President Hayes.)

"At 7:30 line was formed and the entire regiment marched up Friendship Road and down Main Street, and entered Clark's Hall to the music of the band. During the march a beautiful display of fireworks was given. The hall was filled to overflowing. Exercises opened with a band selection – 'Sheridan's Ride' – and continued with prayers, addresses and responses from Miller, Chamberlain, and Butterfield, along with recitations by a local lady, Mrs. Maude Clark Gay. A resolution that thanked the comrades of Company E and the citizens of Waldoboro for the generous entertainment and other kindnesses completed the evening:

'This day, this beautiful village, and your kind hospitality have made this event one long to be remembered. We consider it to be one of the best reunions that the regiment has ever had."

It had been a good day for Waldoboro.

Mr. Samuel Miller – 1973 Waldoboro Bicentennial Book

Somc Storics of Waldoboro's 20th Maine Members

Captain Atherton W. Clark – the most distinguished of the Waldoboro men; in action from Antietam to Appomattox; twice brevetted on the field for conspicuous action in battle; brevet Major October 25, 1864 for gallant service at Peeble's Farm and Hatcher's Run, Virginia; brevet Lieutenant Colonel for gallant conduct before Petersburg.

James H. Stanwood – wounded at Gravelly Run and raised to rank of 1st Lieutentant.

Sergeant Henry F. Sidelinger – raised to Major for meritorious conduct in the Appomattox campaign.

Sergeant Thomas R. Hogue – shot in the thigh at Fisher Hill and promoted to 2nd Lieutenant.

Corporal Raymond W. Hoffses – wounded at The Wilderness and had one foot amputated by a Confederate surgeon.

Corporal Calvin Bates – taken prisoner; lost both feet to exposure in prison.

Private George Benner – wounded in Wilderness Campaign.

Private Lowell Brock – died in a Confederate prison February 1, 1864.

Private Edward K. Chapman – suffered shoulder wound at Bethesda Church.

Private Lewis G. Flanders – died at Andersonville Prison.

Private George A. Hoch – wounded and captured at The Wilderness.

Private Frederick Kinsell – fell at Rappahannock.

Private Leander M. Mink – died of wounds at Augusta, Maine suffered at The Wilderness.

Private Orchard F. Mink – wounded at Gettysburg.

Orrin G. Mink – joined the regiment in 1864 – captured and taken to Andersonville Prison where he was seen alive by Corporal Calvin Bates. Died February 20, 1865.

Sergeant Gardner Schwartz – fell at Fair Play, Maryland.

Private Charles Keizer – fell at Laurel Hill *Note – The G.A.R. Hall in Waldoboro, built well after the War, was named in honor of Private Keizer.

Addendum – According to Samuel F. Miller in his History of Waldoboro, 287 men served the Northern army and navy during the Civil War. (see pp. 143 – 178)

The Depot – Where the 20th Maine veterans arrived
Courtesy Waldoborough Historical Society

Depot Street parade route – Courtesy Waldoborough Historical Society

Memorial Day Parade, 1911 – Courtesy Waldoborough Historical Society

G.A.R. Hall – 1973 Waldoboro Bicentennial Book

Forgotten Buildings (Part 2)
Grand Army of the Republic (G.A.R.) Hall

Number 25 Marble Avenue in Waldoboro marks an unpretentious, white wooden structure at present used for two apartments. It sits quietly down in a gully with precious little parking available around it. (Parking has always been a problem.) Few among us realize that 100 years ago this was one of the most recognized and revered buildings in the village – the home of Waldoboro's Civil War veterans, the Grand Army of the Republic or G.A.R. Hall.

Nearly 300 men from Waldoboro volunteered to join up for the War Between the States – 50 in August 1862 to a three year hitch with Company E of the 20th Maine; 81 with Company A of the 21st Maine and a nine month hitch; 121 to various other units, and 22 to the Navy.

Mr. Samuel L. Miller, owner, editor, and publisher of *The Lincoln County News* and a citizen of Waldoboro, had enlisted in the 20th Maine as an 18 year old private. He rose to be a Sergeant-Major at Gettysburg, then a regimental Quartermaster Sergeant, and finally was discharged in 1865 a 2nd Lieutenant. As a writer and a historian he attended several reunions of the 29th Maine held in various places including Gettysburg and Portland. In 1885 he and 30 of his old comrades pushed to form the Charles K. Keizer Post (named after a member of Company E who died at Laurel Hill, Va.) in Waldoboro. Subsequently, through his editorials, he began to rally the town to build a meeting hall for Civil War veterans. On November 3, 1892 he announced the commencement of a building fund (with the carefully placed exhortation to the notoriously thrifty populace that "There is no need to spend for a STATUE!")

On June 8, 1893, however, he exerted a bit more pressure: "The memory of 400 men who went forth from this town and risked their lives for a united country is still tenderly cherished by their surviving comrades UNMINDFUL of the FORGETFULNESS of an UNGRATEFUL people."

Mr. Miller had his way. The building was erected forthwith, and a dedication was scheduled for August 9. Eight other G.A.R. posts were invited to participate; the Women's Relief Corps of Waldoboro, formed by 29 ladies in 1889, planned to serve 200 meals. It was all very grand: the parade which began at the hall and wound its way down Pleasant Street, around Farrington Corner onto Friendship Street, up to the Brick Schoolhouse and back; Mr. Miller as the Grand Marshal, and the many veterans in their faded uniforms; the speeches; the music; the wonderful food (which gained $120.00 for the ladies). It was a fitting send off to the Charles K. Keizer Post G.A.R. of Waldoboro and its 31 charter members.

For many years the G.A.R. Hall was well used by the aging gentlemen and ladies who had built it. As new wars came and new veterans appeared, they were often invited to the Hall for gatherings and photographs. Finally, when all the old men were gone, the building evolved into the Charles Lilly American Legion Post (named after the first Waldoboro man to die in World War I). About 1973 the Charles Lilly Post moved to its present location on Jefferson Street (better parking!). The old G.A.R. Hall was sold to be turned into apartments.

(Information for this article was gathered from The Waldoboro Bicentennial Book, S.L. Miller's *History of Waldoboro*, *The Lincoln County News*, and conversations with Mr. Richard Wallace, Mr. Stan Hanna, and Mr. Bob Pearson.)

A reunion of veterans (probably from the Civil War, Spanish American War, and World War I.) Only two faces are known: the gentleman seated fifth from the left, second row is John W. Palmer, the longest living veteran of the Civil War. Mr. Palmer, an 8 month P.O.W., died in 1944 at the age of 98. The Navy man, back row on the right is Josiah Jameson, grandfather of current Waldoboro resident, Galen Harkins. Photograph courtesy of Mr. Richard Wallace.

Lincoln County's Last Survivor

Waldoboro's Civil War veterans numbered 287 – more than one tenth of the town's population in 1860. When Mr. Lincoln called for reinforcements, as he did several times, Maine's counties were assigned quotas; Waldoboro was, in turn, asked to supply a certain number of men. As time went on, and the "glory of war" found not to be all that glorious, the threat of conscription or draft was invoked in order to encourage volunteerism. Money was raised as incentive also.

At first there was no rush to form an official Waldoboro company. Men enlisted in the other six Maine regiments – particularly the 4th Maine which had its rendezvous in Rockland in June of 1861. The first man in town to join was Jacob C. Bogue, enlisting April 17, 1861 – less than a week after Fort Sumter, and there were a few others. But the truth held that there was an element of anti war sentiment in town – encouraged by some – not all – of the powerful and influential shipbuilders who feared the loss of their cotton trade to Europe.

1862 saw a clear call for 50 Waldoboro men to join 30 from Bristol and 20 from Union. A $100.00 singing bonus was offered to which some private citizens added $650.00 to be divided among the enlistees. Company E of the 20th Maine regiment was born, soon to participate in many battles, including under Joshua Chamberlain at the famous defense of Gettysburg's Little Round Top. The war ended, and on June 17th, 1865 11 survivors of this company trudged together down Main Street to be honored at a public reception in Union Hall (Fish Block).

Almost immediately after Company E had departed for war in 1862 another call for men came from Washington. Lincoln County was to supply 411 – Waldoboro 90 of these. On August 23, patriotic music from the local cornet band and speeches filled the air outside the Customs House (current library). Townsfolk voted $5,000 (immediately augmented by two local bank's loans of $4500 each) to entice enough young men to form the needed three month group of enlistees – Company A of the 21st Maine. The enticements were successful, and Company A, led by local hero Isaac Comery, left to play an important role in the Mississippi River battles of Port Hudson and Vicksburg.

From the 457 Waldoboro men who served these two and other units, 37 died and 26 were wounded. 14 were given or earned commissions. Many also served time in the infamous Southern prison camps, and only six of these men survived. One of them was John W. Palmer, not actually born in Waldoboro but who lived the last 76 years of his 98 year life here and was a stalwart member of the Charles Keizer G.A.R. Post until his death in 1944.

From the Waldoboro Bicentennial Book – "John W. Palmer was the last Civil War veteran in Lincoln County. Mr. Palmer was born in Nobleboro on June 22, 1846. In February 1863 at the age of 17 he enlisted in Company I 32nd Maine Volunteers. To reach Bath he walked down the East Neck Road, where he lived, to meet the stage. Of the 13 boys from that section who fought in the war he was the last survivor.

His regiment fought in Spottsylvania, Cold Harbor, and the Wilderness. Badly cut to pieces in Petersburg, Virginia, when the regiment was caught and trapped in a mine crater on June 30, 1864, Mr. Palmer was one of the few survivors. He was taken prisoner and confined for a month at Libby Prison at Richmond, Va. He was transferred to Danville where he was held for seven months. He was discharged from the service July 15, 1865. Three years he moved to Waldoboro."

In a newspaper interview in 1939 when he was 93, Mr. Palmer remembered the following about his experience in prison:

"We were given a little corn bread once a day. We would eat that up and go without food until the next morning. When we entered the prison, it was pretty hot, and we would take off our clothes and lay them on the hard pine floor. Then we would lie on our clothes to sleep. I had a brick, an old pair of shoes, and a rebel cap for a pillow. When it got cold, we all slept spoon fashion along the hall, so crowded that when we wanted to turn, the man at the head would say 'Spoon', and we all turned."

(Material for this article was procured from Stahl's and Miller's Histories of Waldoboro and the Waldoboro Bicentennial Book.)

Mr. John W. Palmer
1973 Waldoboro
Bicentennial Book

A Grand Lady of Waldoboro – Maude Clark Gay

In late August of 1896, one hundred surviving members of the famous Civil War 20th Maine regiment held a never-to-be forgotten reunion in Waldoboro. Arriving by rail, horseback and carriage, they marched into town and met formally with commander Joshua Chamberlain at the new G.A.R. Hall on Marble Avenue. These comrades from three years of war – including the Gettysburg campaign – spent the afternoon talking and catching up, listening to a Waldoboro Cornet band concert on the bandstand across from the Customs House, and touring the busy village. After supper General Daniel Butterfield led a parade of soldiers and musicians down over the Main Street hill to Clark's Hall where a throng of guests and townsfolk awaited them. A full program of band selections, patriotic poetry and prose was followed by prayers and addresses from regiment veterans. A *Lincoln County News* report (modestly) summarized: "It had been a good day for Waldoboro."

One of the speakers that evening was twenty year old Maude Clark Gay who gave several "recitations". The concept of a woman speaking before such an assemblage was unusual (if not unheard of) in these times; however, Maude Clark Gay had for several years been recognized as a talent and a leading figure – despite her youth. Her lineage, of course, was impeccable – granddaughter of Civil War hero Colonel Atherton Clark and great granddaughter of one of Maine's greatest shipbuilders, Joseph Clark. But Maude Clark had accomplished much on her own: a high ranking student and public speaker in the schools of Waldoboro and Lincoln Academy in Newcastle, a thespian on local stages, it was no surprise to have her featured in the hall built by her great grandfather. Hers and the other speeches as well as the musical selections were beautifully rendered; the evening was a complete success.

Maude Clark Gay had recently been wed to John T. Gay, Jr. who, with his father owned one of the large grocery stores advertising "groceries, corn, flour, provisions, paints, oils, and brushes". Mr. Gay would run this establishment until his death in 1937; she would outlive him by 15 years. One child was born to them – Anne Gracia. They lived in a large house on upper Main Street formerly owned by News publisher and historian Samuel Miller.

In the years following the 20th Maine reunion, Maude Clark Gay continued to demonstrate her individualism, her talents, and her courage in a "man's world". *Lincoln County News* reports of March 1899 find her reciting poetry ("Winning Cup's Race") at a public gathering in the Masonic Hall and later that month a two person (with Alice Reed) rendering of Shakespeare's Macbeth Act V, Scene 1 at Clark's Hall. During World War I she acted as Secretary of the local Red Cross chapter. She was a long time stalwart member of the Woman's Club starting at age 18, eventually becoming president of it and also of its affiliate, the Lincoln County Union of Woman's Clubs. In later years she traveled to many other states and spoke as Maine's Woman's Club representative.

Politically, she ran for and served in the Maine House of Representatives from 1927 to 1930 and in the State Senate 1933 – 1934. Finally, she holds the distinction of being the first woman to cast a vote in Waldoboro. On the day in 1919 when the state polls were finally open to women, Maude Clark Gay took her stand at the ballot box and waited – long before anyone else – male or female – had arrived.

For the first time in Waldoboro's long history a woman had been allowed to cast a vote! (Author's note – How far we have come; but can anyone imagine Maude Clark Gay being stopped from voting that day – or any day thereafter?)

Maude Gay also became a well known writer of poetry, short stories, magazine and newspaper articles and two full length novels. One of her short stories (concerning a romance in a Maine seacoast village) will be our next entry.

Sources for this story include *The Lincoln County News* and Stahl's *History of Waldoboro*.

Maude Clark Gay
1973 Waldoborough Bicentennial Book

Maude Clark Gay – Grand Lady of Waldoboro
Part 2

Maude Clark Gay (1876–1952) was a Waldoboro civic leader, a pioneer in equal rights for women, and an elected politician, serving in both the State Senate and the House of Representatives. She was also an author of two novels, poetry, and short stories, many of which appeared in magazines and newspapers. The following is an example of her work.

"The Robin – A Tale of Old Winterport"

They called him hard hearted down there in the little fishing village. Stern and cold, masterful and domineering, and a half dozen other names they called him. If he knew, he gave no sign but went his way, heedless of comment or criticism. As was the custom with the sons of Winterport, he answered the call of the sea. During his mother's life, he seldom left the great house on the hill when at home from voyages. The old servant, who had been his nurse, was perhaps the only one besides his mother who saw any traits besides harshness and austerity in Miles Farrington's nature. After he had followed his mother to her last resting place, this old servant whispered to an interested neighbor that "Master Miles was a – grieving sore." If he did grieve for the mother who adored him, he gave no outward sign. Soon the sea claimed him for his next long voyage to foreign lands.

On his return two years later he announced his intention to retire from seafaring life. The surprised people of the village decided that he must have made a "handy pile of money" and wondered what he would do with it living all alone in his great house on the hill. The last thing the village people expected Miles Farrington to do was precisely what he did do. He went down the hill to the little cottage on the shore to "pay attention" to Widow Lewis' daughter Charmian, the pride and the beauty of the village.

A wave of indignation swept over Winterport. "Miles Farrington to even THINK of marrying Charmian! That crusty old bach," whispered the younger generation. "It is a SIN! Guess any girl would put in a life of it with HIM!" announced the elders, sagely. "Her mother was a lightheaded, frivolous thing!" croaked the gossips. "If she hadn't been, she would NEVER have called that girl Charmian!" But even THAT maligned old woman, now an invalid, shrank inwardly when one day she saw the darkbrowed captain turn up the path to her door.

Charmian herself shrank from the man, although a strange thrill ran through her sometimes at the earnest look in his steady eyes. "If Miles were not so hard hearted," she would say to her mother, "If only I could see one tender thing about him," and again wistfully, "How his mother loved him; he was so kind to her."

Winter came quickly and closed in on the little town. Work was slack, and many a ship carpenter worried about his supply of wood and coal. And just then Miles Farrington announced that he would build a vessel. This caused great excitement in the village. Work for Captain Miles meant steady pay. After the frame of the brig was cut in the woods beyond the village, the keep was stretched down to the shore. From the windows of Charmian's house the sturdy figure of Captain Miles could be seen moving about his men. If the girl paused now and then at her household tasks to linger by the window, no one knew it but her mother, and she never told.

All through the sparkling winter work progressed on the brig, and plenty reigned in the homes of the sturdy carpenters of Winterport. When the ice drifted from the Penobscot, the vessel was all planked, and the crew was laying the decks and building the houses. When the soft breath of April stole on the air, the riggers came down from Bangor, and the masts of the brig began to rise against the soft spring sky. When the first May blossoms pushed through the moss in the dim woods, the work was done. The sturdy little brig with its full bows stood all ready on the stocks, an honor to the builders of Winterport. As vessels went in those days "she was a beauty", and people came from near and far to see and admire.

One fine Sabbath morning Charmian Lewis walked down the quiet streets with Captain Mile to see his new vessel. The gossips, peering through their windows, shook their heads and wondered "if he had asked her yet, and if she would be fool enough to have him." As they stood on the high bank, where the jibboom reached in among the trees, Charmian cried out, "Oh, look, Captain, look! There is a bird's nest on the vessel." True enough, there was the next, tiny and frail, snuggled up under the bowsprit, where the stick lifted a little from the figurehead. The mother bird had just brought a fat worm to her family, and faint peepings could be heard from the little nest. Charmian turned to the captain with lovely, distressed eyes.

"Oh, the launching!" she cried with a little sob, "Oh, the baby robins! What will become of them?"

"The woods are full of robins," answered Captain Miles, turning brusquely away.

A queer little pang smote that girl's heart-strings as she looked at his stern face. She did not visit the vessel again. A week stole by. Two weeks were gone and no date set for the launching. No one dared to question Captain Miles. The village people whispered among themselves, wondering "if he was keeping her just to look at." Freights were high in those days, and vessels for the West Indies were in great demand in Bangor. As the captain was often seen looking down at the figurehead lost in thought, the word began to be rumored about that "Captain Miles seemed wrong in his head" and folks wondered if Charmian had refused him.

The May sun mounted in the sky, and the new paint on the sides of the brig began to show in ugly little blisters. "Taint doing her any good!" the old ships carpenters declared, as the days grew into the sultry languor of June. "One of his odd notions, queer as well as hard. Why, he hasn't even given her a name."

It was true. Every vessel built in Winterport had been named long before being ready for sea, but this vessel had a stern whose blankness did not appease one whit the curiosity of the village people. The mystery of the unnamed, unlaunched vessel had almost ceased to be discussed as the June roses drooped in the hot sun.

One night, when the fireflies flitted in the tiny garden, Charmian Lewis sat alone on the steps, a white slip of a girl in the soft dusk. A big, stalwart figure came down the street and paused by the gate.

"I know you aren't interested in me," Captain Miles said slowly, "But you will be in what I'm going to say. The robins are big enough to take care of themselves. We'll have the launching tomorrow."
For a moment Charmian sat in silence. The summer darkness closed in about them. A great light was quivering and shining in the girl's soul. "You did not have the launching, because of – the robins – the little nest," she said brokenly.

"My mother loved robins," answered the man's steady voice, deep with the undertone of some stirred emotion. "She watched for them in the spring and put out food and bits of cloth for their nests. I always thought they missed her somehow after she went away. I would have let that vessel rot on the stocks before I turned those baby birds out of their home. But they can fly now. We'll be launching tomorrow."
It was a long speech for the reserved, taciturn man. When he had finished, there was another throbbing silence. The girl drew in her breath with a quivering sob. Her lashes were wet with tears.
"Miles," she said, her soft hands on his rough sleeves, "I never dreamed that you could be so tender. Please forgive me, Miles."

Miles Farrington must have forgiven her, for when the vessel slid down the ways in the glory of the June morning, Charmian Lewis stood by the captain's side and christened her "Robin".

Good luck followed the Robin through many years on long voyages to foreign lands. She brought back gold, too, to fill the coffers of Captain Miles and Charmian, his lovely bride, in their great house on the hill. And, after many years of usefulness, the Robin came back to lie at rest on the white shores of Penobscot Bay. The robins sing in the springtime from the same elm tree whose branches wave over the mouldering limbs of the brig that showed a man his mind and a woman her heart and led them both to ways of happiness and peace.

The source for this writing was the booklet "Echoes From Old Doorways – Little Romances of Early Maine" by Maude Clark Gay – printed and bound by the Journal Printshop and Bindery, Lewiston, Maine

A Celebration – With Problems

On June 28, 1773 the General Court of Massachusetts granted an Act of Incorporation to the Town of Waldoboro in the District of Maine. One hundred years later a centennial celebration was observed; after a public meeting indicated much support for the idea, 46 townspeople formed committees of Management, Subscription, Correspondence, Invitations, Address, Dinner, Parade, Decorations and Motto, Fireworks, Setting up the Public Service, and Finances. (Principal citizens contributed liberally to the latter, making this project possible to undertake.)

As the day of incorporation, June 29, fell upon a Sunday, the Management Committee decided that July 4 would be more appropriate. Smouse's Grove (the large field south and east of Kaler's Corner) was selected as the place for the public service. There a fine platform was erected for the speaker, and spacious tables were constructed for the dinner.

At sunrise on the 4th a salute of 38 (number of U.S. states) guns from Prock's Ridge high above Smouse's Grove announced the day. In the village the streets began to fill with Waldoboro folks and thousands of strangers who had arrived via every conveyance including special trains from east and west. Flags floated in the breeze from dwellings, stores, and from ships anchored in the harbor. Martial music from groups rehearsing filled the air.

11 A.M. – a procession began to form at Farrington's Corner (Upper Main Street and Old Route One). At 12 noon the marchers stepped out, heading down Main Street toward the business area. In order they moved: chief marshals and aides; The Waldoboro Cornet Band; 40 soldiers dressed in Continental garb; the orator of the day; Colonel A.W. Bradbury of Portland, in a carriage; municipal authorities; 14 aged (over 70) citizens in carriages – including Mr. James Light, aged 101; a large boat with 38 young ladies riding in it; The Damariscotta Cornet Band; fire companies and highly decorated with bunting fire engines from Waldoboro and Thomaston; another large boat with 13 young boys representing the Navy riding in it; The Goshen Drum Corps; Templars from three sections of town; A company of mechanics carrying tools; wagons representing J. Clark and Son, the Waldoboro Suspender Co., H.M. Folsom Co.; a Sprague Mowing machine representing Richards and Storer; and finally a large group of public citizens. The procession was so long that as the head of it was turning into Jefferson Street, the tail was still at Farrington's Corner!

But now disaster struck the happy day. Just as the last group of walkers stepped forth, a shout of "FIRE!" was heard. Mr. Newell Winslow (leader of the cornet band) was informed that his house was burning. The procession halted, and a valiant effort was made to press the fire companies and engines into action. Unfortunately, most of the equipment was bound up with bunting and flags, and the men were unable to disentangle it in time to make a difference in the saving of the Winslow home. Confusion reigned supreme for about an hour as firemen fought to prevent the flames from spreading. Successful in this they were; sadly, they received little help from the spectators who chose not to soil their holiday clothes and ignored the exhaustion of the firefighters. One of the latter group, a Charles Chase, was overcome by the heat; another, Edwin Clark, was struck by a falling section of hose and injured. (Both men recovered after a few days of rest.)

At 2 P.M. with the fire finally under control, thc proccssion was reformed and continued its route – down to Jefferson, right turn, over to Mill Street, left turn, across the Soule Bridge to the present Route 32, left turn, and then up to Smouse's Grove. There at the designated area, 600 hungry people enjoyed a "sumptuous collation". Speeches (Mr. Bradbury's took an hour to read), poems, and music by the band followed, the original Act of Incorporation was read, and finally the throng joined in with the singing of "America". The parade reformed marched down into town, and disbanded late in the afternoon.

Evening of July 4, 1873 brought a series of rain showers which effectively dampened hopes for a large display of fireworks, and only about one half of the supply was used. As Mr. Samuel Miller, an eye witness, reported, however, "with the exception of the fire, the day had passed off without disturbance of any kind, and it is a credit to Waldoboro that such an occasion was celebrated without the usual accompaniment of drunkenness and rowdyism."

Sources for this story were Miller's and Stahl's Histories of Waldoboro plus a newspaper article from the *Waldoboro Press* – June 29, 1967 – and given to the writer by Mr. J. Malcolm Barter of Dutch Neck. Many thanks to Mr. Barter.

Waldoboro Cornet Band – June 22, 1885 – 1973 Waldoboro Bicentennial Book
Back row: Granville Waltz, James Stanwood, George W. Welt, Joseph D. Johnson, Herman Kuhn, Fred D. Redman, I. Gardner Reed, leader
Seated: Joseph Simmons, Leavitt L. Lambert, Jefferson P. Vannah

The Lincoln County News – June 26, 1874 – "The Waldoboro Cornet Band-16 pieces – will furnish music for levees, picnics, excursions, political gatherings, military parades, etc. at short notice and at reasonable rates. Apply to H.N. Winslow."

The Depot
Author's Collection

The railroad comes to Waldoboro in 1871. Villagers voted that the Depot be placed "3/4 of a mile from the four corners".

The Lincoln County News – July 3, 1874 – "Trains – Waldoboro for Thomaston – 7; 8:45; 10:48 A.M.; 3:08; 5:07; 7:10 P.M.
Thomaston for Warren and Waldoboro – 6:15; 10:10 A.M.; 12:45; 2:30; 6:15; 10:45 P.M.
Round Trip Price – 75 cents. Tickets good on any train during the day."

The Village at its Grandest – 1884

After the terrible fire of 1854 much of Waldoboro Village was rebuilt and revitalized – thanks in great part to profits from shipbuilding. A study of an 1884 town map from a collection found at the historical society shows clearly what a busy and thriving place it must have been.

A traveler, journeying to Waldoboro from east to west would move down the steep, dirt packed hill to the Main Street bridge, acutely aware of the still busy shipyards to the south. Sounds of pinging caulking hammers, steam driven saws, and the sight and sounds of perhaps 200, or more, busy workers were introductions to the town. Just after crossing the bridge, he would note a two story building on the left which housed general storage on the ground floor and a sail loft belonging to Stephen Jones on the second. Moving up Main Street and taking a left turn (below the area of the present day Grange Hall), he would be on Mill Street – a driveway today. A few hundred yards down this street he would come upon two large mills. On the right stood the James Boyd Iron Foundry (coal and steam driven), manufacturer of large iron implements for ship and farm use. Further along, on the river bank, the remains of its dam still visible today, was B.B. Haskell's Grist, and Planing Mill. Back along Mill Street were sheds, storage buildings, and three more businesses – a printer, a wheelwright, and a blacksmith. As he moved up the Main Street hill, the traveler could view on the left (present day Grange Hall parking lot) the 30 room white wooden Medomak House Hotel with its large livery stable and carriage house. A driveway separated these buildings from a vacant structure and a meat market. Next door stood the Jackson brick block (present day dentist's office) which was at the time occupied by a photographer and a crockery shop. Finally, on the corner of Main and Jefferson Streets stood the famous Fish (or Union) Block – "the finest business block in Lincoln County" – five stories in the back, four on the street, and three in front – housing the Waldoboro Exchange Hotel, a drug store, a clothing store, and two grocery shops.

The sightseeing traveler crosses Main Street at this point and moves back down the hill toward the river. The brick block on the corner of Friendship Street and Main was then called the Hovey – Genthner building, and the corner was Balch's Corner. A narrow alley separated this structure from the famous Clark's Hall – a wooden structure which housed a grocer and a tailor on the first floor and a much used (nearly every night of the week – for drama, music, and lectures) theater on the second. This theater, incidentally, was lighted and heated with gas – a new and very modern idea when the hall was built in the 1870's. Connected by a wooden bridge to Clark's Hall was the next building, also owned by the shipbuilding Clark family, which offered a tailor shop and a candy and cigar store on the first floor, an apartment on the second, and the Masonic Hall on the third. Next on the west bound downward slope was a vacant building, then another meat market, and then, set back from the street, a large livery stable for the Waldoboro Exchange Hotel. Another small structure with a printing office in the basement and a carpeter on the first floor followed by a two family home completed the Main Street dwellings as the traveler descended toward the bridge.

Back up at Balch's Corner and on Friendship Road, the outsider's view (besides the unpaved street) would not be at all that different from today's. The 1884 traveler would see the same brick blocks, all built in 1855. In these blocks, as he moved east were, on the right: a millinery shop, the express office, the Odd Fellows Hall, two groceries, stationery and "fancy goods", a stove shop, dentist's office, a jeweler, a harness shop, and a livery stable. Across Friendship Road, in the Haskell (later Sproul) Block were a millinery, grocery, bank, hardware, tailor, billiards, fish market, and a barber shop. Across Pleasant Street (then Dog Lane) and continuing on down Friendship was a large grocery and finally a fruit market.

As the 1884 traveler prepared to leave Waldoboro and moved back down Main Street, he would notice, west of the Friendship Road business area and in the present day parking lot space, a large building with the sign – "E.O. Clark – Saw and Planing Mill for Ship Work." Unbeknownst to the traveler, this mill had been recently closed – a sign of sad future events having to do with shipbuilding in Waldoboro.

Village merchants sweeping Friendship Street
1973 Waldoboro Bicentennial Book

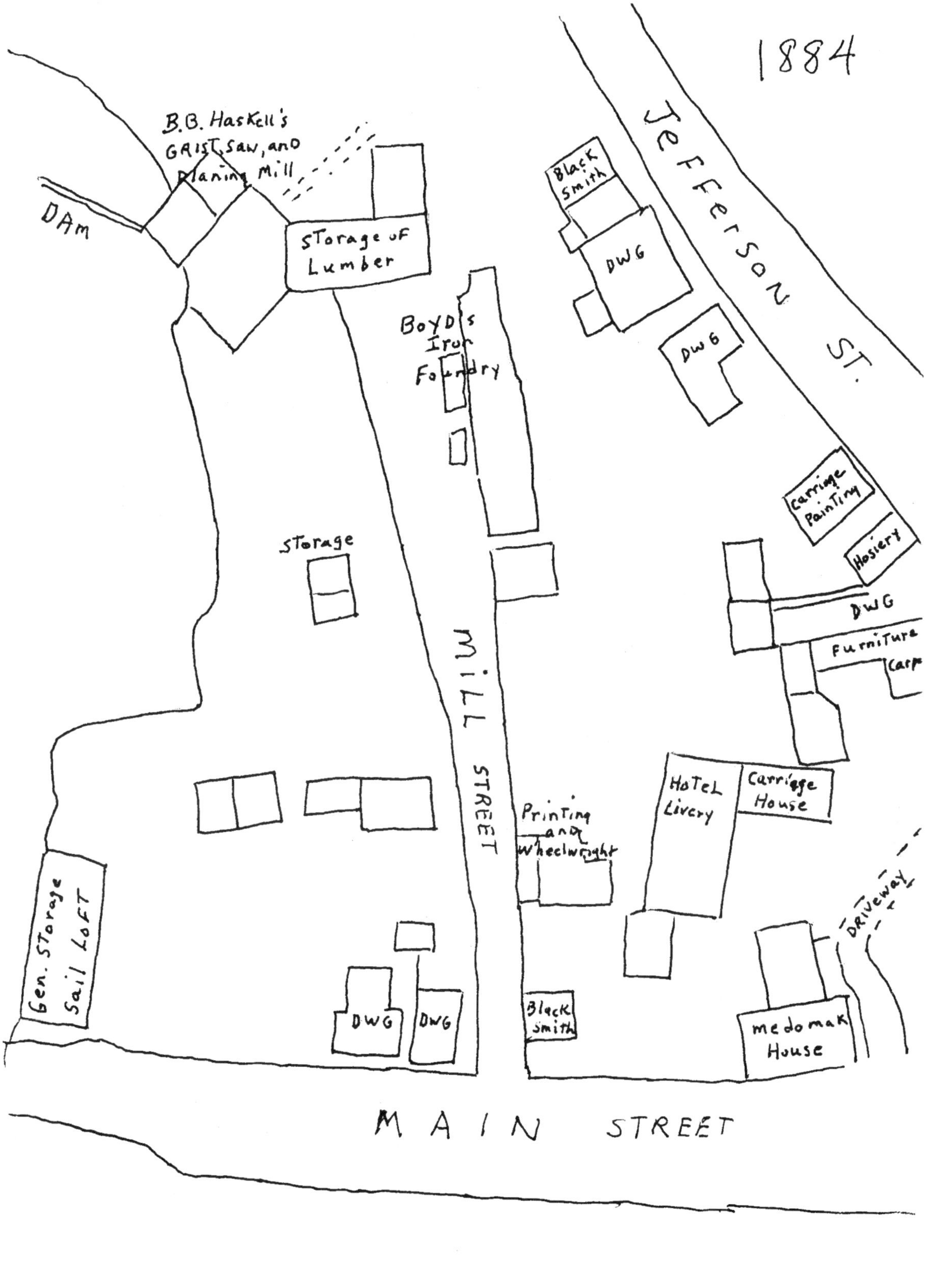
1884
B.B. Haskell's
GRIST, SAW, AND
Planing Mill
DAM
Storage of
Lumber
Black
Smith
DWG
DWG
JEFFERSON ST.
Boyd's
Iron
Foundry
Carriage
Painting
Hosiery
Storage
DWG
Furniture
MILL STREET
Hotel
Livery
Carriage
House
Printing
and
Wheelwright
Driveway
Gen. Storage
Sail Loft
DWG
DWG
Black
Smith
Medomak
House
MAIN STREET

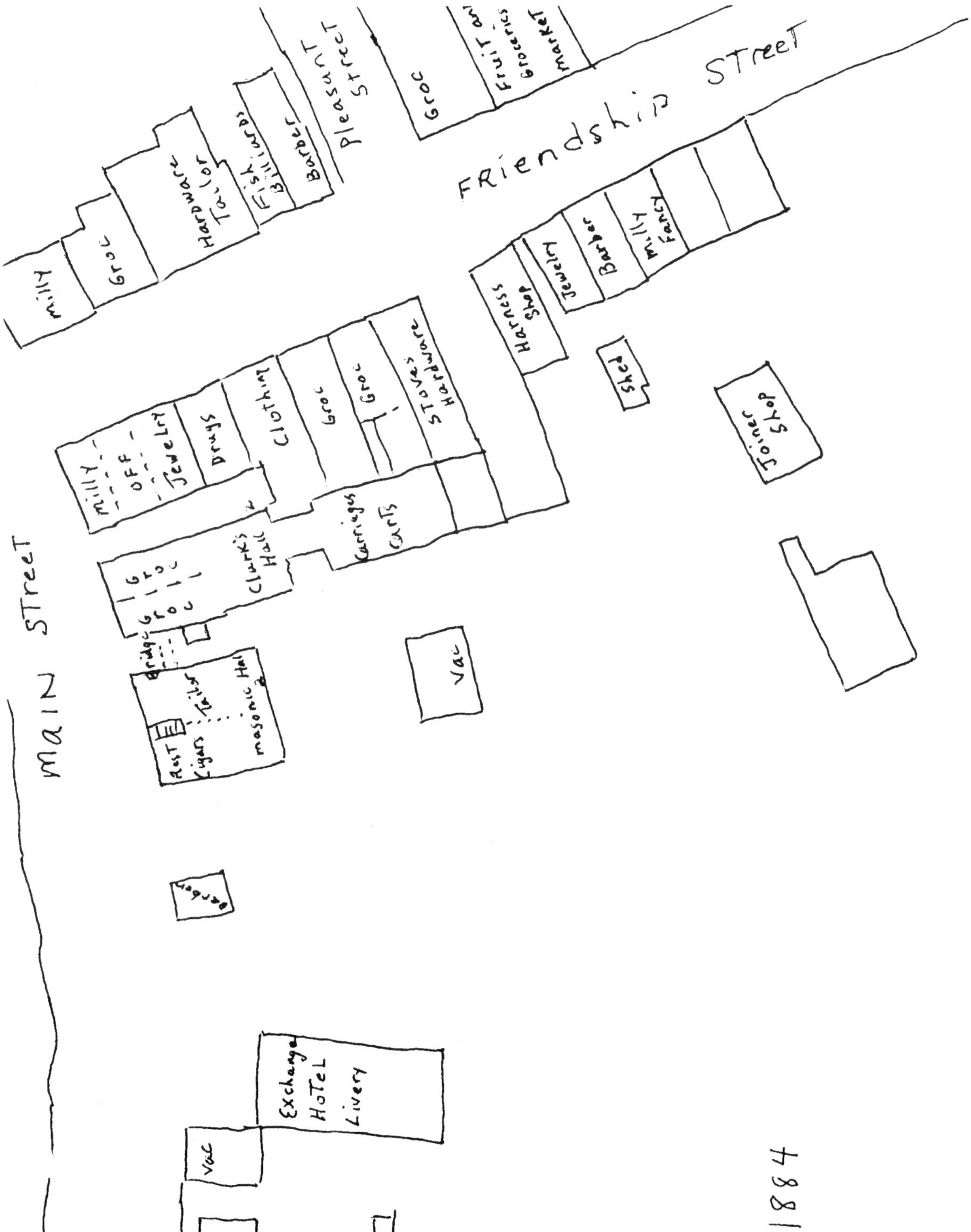
FRIENDSHIP STREET
PLEASANT STREET
MAIN STREET
milly
Groc
Hardware
Tailor
Fish
Billiards
Barber
Groc
Fruit and Groceries
market
Harness Shop
Jewelry
Barber
milly
Fancy
Shed
Joiner Shop
milly
OFF
Jewelry
Drugs
Clothing
Groc
Groc
Stoves
Hardware
Carriages
Carts
Clark's Hall
Groc
Groc
Bridge
Tailor
masonic Hall
Rest
Cigars
Vac
Exchange
Hotel
Livery
Vac
DWGS
1884

Hotels of Old Waldoboro

Remembering that travel on land was slow and difficult – much more so than travel on water – it is easy to understand the need for available overnight lodging. Waldoboro, a thriving community of businesses throughout much of its history, attracted many travelers – shipyard workers too far away from home to commute, salesmen, shoppers, itinerant entertainers. Local friends occasionally needed a convivial place to spend part of an evening after a play, concert, or lecture at Clark's Hall, a sleighing or buggy ride, or simply a get-together. Hotels in the great shipbuilding days were also booked by out of town ship owners for gala dinners on the eve of a launching. Over the years there have been a number of such places in Waldoboro.

Jasper Stahl, the great historian, speaks of the early "ordinaries" or taverns – considered social responsibilities for every 18th century town where "neighbors and travelers met, talked, drank, and lodged." Taverns in those days in Waldoboro included Ezekiel and Mary Barnard's on Main Street, also a stage stop; Sampson's on Thomas Hill in South Waldoboro; "Aunt Polly Klaus" across the road from the German burying ground; and "Aunt Lydia's" out on the road to Warren (now Route 1).

After the great fire of 1854 many new buildings appeared in town. Among them was the Medomak House, a 30 room hotel with a large stable, built approximately where the parking lot of the Grange Hall is today. Owned by the Miller family for many years, it was a white, clapboard structure and boasted the stagecoach stop as well as providing, in the 1870's and '80's "easy service by hack to the depot." The Medomak House burned in 1892.

On Friendship Road in the building now housing a laundramat and upstairs apartments, stood another hotel known by different names – "Lincoln" in the 1850's and again in the 1920's – and "Savoy" in the '90's and early 1900's. A *Lincoln County News* entry of January 18, 1900 describes "A lively group of young people drove over from Damariscotta Tuesday evening and partook of Landlord Trowbridge's fine supper at Hotel Savoy. After an hour or two of social engagement they departed." Stan Hanna, long time Waldoboro resident, recalls with pleasure the jovial Irishman, James Dwayne, who owned the Lincoln Hotel (along with another hotel in Camden).

Street maps of 1884, 1891, and 1896 show a third hotel – the "Exchange" – later "Windsor". This hotel was found in the Fish Block, an imposing five story brick structure on the corner of Main and Jefferson Streets. The livery stable for this establishment was located across Main Street in the current parking lot. This Exchange – Windsor edifice burned along with the other half of the Fish Block in 1900.

1899 saw yet another hotel, the "Davis" open in the old Masonic Building adjacent to Clark's Hall on the south side of Main Street. Mr. Davis carried out his plans for an office 20 feet by 34 with lavatory on the first floor along with a kitchen, dining rooms, a parlor and other sleeping rooms – 16 in all – on the second. This building was the place that William F. Palmer, owner of the famous Palmer fleet designated for his large Boston parties when launchings of his six schooners occurred. In a letter to his local builder, George Welt, September 24, 1901, he rather imperiously ordered: "Please engage the entire Hotel Davis, including the dining room, for our breakfast and a quick lunch after the vessel (Baker Palmer) is in the water."

Finally, many old timers can remember the establishment known as "Stahl's Tavern", currently the home of Ted and Betsy Wooster. The tavern was once owned and operated by Clint and Ida Stahl and could house about 15 guests. For 40 years resident Louis "Bushy" Boissoneault was employed there as chef. According to Bushy, there is no truth to the rumor that General Dwight D. Eisenhower and singer Bing Crosby were guests (although Mrs. Stahl did once appear on Crosby's show in New York). Martha Boggs remembers the wonderful ice cream served at the tavern and also taking the Greyhound Bus from outside its doors to Rockland during World War II when gasoline was scarce.

From Barnard's Tavern and the stage stop at Medomak House to the bus stop at Stahl's Tavern, hotels have played an interesting and important part in Waldoboro history.

Hotel Lincoln, once "Savoy", and in this photograph "Waldoboro"
Author's Collection

Waldoboro, Maine

*Believed to be photo (circa 1870) of the four Winslow brothers, sons of Newell and Mary Jane (Waltz) Winslow of Waldoboro. Plate on caps reads "Winslow Hotel".**

Back sitting: Willis (Willie) Stone Winslow (1857-1899).
Front l to r: Leroy (1853 - ?), Harland Newell (1849 - 1913), and
Samuel Snow Winslow (1851 - 1920).

**From the collection of the late Annie K. Winslow Golding.*

Stahl's Tavern – The builder, John Brown, once entertained his friend, President John Adams, at his previous Waldoboro Home.
Author's Collection

Stan Bailey Reminisces

I. Stanley Bailey, known always as Stan, was born in 1902. Growing up, times were hard in Waldoboro, and the family moved several times as father worked at different jobs – including chicken farming. Stan graduated from the old Waldoboro High School, now Friendship Street School, went on to the University of Maine and immediately after college to the carpet business in Massachusetts. Soon he was to join another carpet company, the Lee Company, for whom he would work for many years as a very successful salesman. After World War II he and a friend tried the import/export carpet field which involved a good deal of travel (including to Afganistan). Tired of long trips, he retired and came home; retirement for Stan, however, was hardly to be inactive. Extremely loyal to his roots: Waldoboro town organizations (Lions Club, Library, Historical Society), the University of Maine, and old friends, he remained an energetic leading figure well into his 90's. Winters were spent in Florida (with his own car transported by rail), but every spring he and his wife returned to their house on Friendship Street full of enthusiasm and cheer for Waldoboro and her people. It seemed sometimes that Stan had discovered the Fountain of Youth and was going to live forever. But then the news suddenly came back from Florida in the winter of 2001 that he had passed away. For his incredible long lasting history of friendship, generosity, and sound advice, he is sorely missed by the many who knew and loved him.

In November of 1999 I had asked if I might persuade him to reminisce a bit about his youth in Waldoboro. Here are some of his recollections:

Stan had worked at different Waldoboro establishments while in school and college. At the Star Theater just west of Balch's Corner and adjacent to the current "Purple Foot" there was a bowling alley where he toiled as a pin setter, earning 2 cents a string for his efforts. He handcranked the movie projector at the Star; he ran a roasted peanut business across Main Street at the Gay grocery block. (Apparently the smell of peanuts was irresistible to shoppers – especially in the summer time!) He cut cloth patterns for shirt collars at the clothing factory (currently New England 800), and he served as a clerk in a grocery store.

It was while working in the theater that Stan met a girl Ann Clark Gay (she played the piano for background music of the silent movies and took tickets) who eventually was to become his first wife. He remembered getting into trouble with Ann when he and another boy substituted charcoal for the white face powder that she always applied to her face after returning from lunch. She eventually forgave him!

In the clothing factory an electric motor in the basement was harnessed by a huge leather belt to the cutting machines on the second floor. Stan and other workers would draw patterns for clothes such as men's collars on a stack of many pieces of cloth; then the boss in the basement (sometimes fortified by a swig from a bottle he kept hidden in the wall) would begin the cutting process. Women and girls, as well as men and boys were employed.

In the grocery store Stan sold everything – including once some vanilla extract and denatured alcohol to a pair of individuals needing a substitute for the liquor whose sales was banned for a time in Waldoboro. (These individuals became to their sorrow exceedingly ill after ingesting the stuff!)

One very snowy night Stan had just finished his duties at the theater and had started to walk home to the family residence on Depot Street. The snow was deep – there were almost no automobiles owned in Waldoboro, so no plowing had been done. (In fact, after a snowstorm, each house owner in and around the Village was expected to shovel the wooden sidewalks adjacent to his home.) As he trudged along, he suddenly noticed two large, yellow, glowing eyes approaching him on his side of the street. Fearing that it might be a wolf or a large coyote, he turned and hurried back to the theater. A friend volunteered to go home, pick up his shotgun, and escort young Stan all the way to Depot Street – well beyond the area of the present Route 1. To the great relief of the two companions, the creature – whatever it was – had disappeared.

Main Street – pre 1890 – Note Medomak Hotel (30 rooms) – white building above Ward Ink sign; Union or Fish Block – large brick building top middle – Author's Collection

The Lincoln County News – May 9, 1901 – "Have patience, Main Street folks. Lumber for the sidewalks is coming!"

Waldoboro Mills
Author's Collection

As early as 1800 there were 13 mills in the town. The three waterfalls in the village were all harnessed along with many other brooks and streams.

Mills of the Mill Streets

Waldoboro's manufacturing history is dominated of course by shipbuilding. Shipbuilding, however, was only one of the industries to find roots here. Mr. Samuel L. Miller in his History of Waldoboro mentions that as early as 1795 there were seven saw mills and three grist mills in operation. In this article we shall take an imaginary walk along the Medomak River and follow it from the "Upper Falls" down through the village past the "Middle Falls" and end up at the "Lower Falls", a short way from the Main Street bridge. We will learn on our journey what industries flourished at these important places of natural power through the years.

We begin at the corner of Mill Street and Bremen Road and walk down Mill Street toward the river. For information we are depending upon several sources: street maps of Waldoboro in 1857, 1884, 1891, 1896, 1904, and 1927, as well as accounts by Jasper Stahl, Bill Bunting, Ida Winchenbach, Sherman Vannah, and Geraldyn Porter. (First, however, it is necessary to state that the term "Mill Street" is somewhat confusing to the Waldoboro historian. The original street by that name was located just west of the current Meenahga Grange Hall on Main Street and led down to the river from there. Today it is a driveway. The modern Mill Street was originally called Jefferson Road.)

At the corner of Mill Street and Bremen Road, then, we start, with Benny's Fish Plant on our right. In the late 1800's the Twitchell-Chapman corn and peas canning company employing about 20 (mostly girls) and open in late summer – stood where Benny's stands. Directly across the street on the river bank there was in 1857 a Soap Factory. Walking along Mill Street toward the current V.F.W. Hall, we come to the "Upper Falls", a place where much activity occurred. In 1877 an ancient grist mill was replaced by Rufus Achorn's imposing Medomak Flour and Saw Mill. The 1904 map shows an Electric Light Plant in the same spot at the Upper Falls, and later Philip Cohen's Poultry Processing Plant stood here.

The Mill Street bridge over the river is called by old timers the "Soule Bridge". In 1857 the firm of Soule and Ewell operated a carding mill on the west bank and an oakum mill on the east bank. (Mr. Soule's house is still standing in the trees just before the bridge.) By 1900 these industries had been replaced the Waldoboro Steam Laundry (east) and G.O. Waltz, Woodworker (west). This is the Middle Falls.

Before turning right on Jefferson Street we look left. Just north of the modern Elm Street (once called Lovell Street) was found in 1857 the Savage and Stanwood Pottery Shop and just north of the bridge a Tin Making Shop.

As we make our way along Jefferson Street toward the village area, we try to smell the familiar coal fires built and used by the several blacksmiths who plied their trade in Waldoboro. On Jefferson Street alone there were two in 1884, and the names – White, Waltz, McIntyre and Johnson were well known. At this point in our journey we move diagonally toward the Medomak and arrive at our final stopping place, the Lower Falls section of town.

Here was a very busy manufacturing district – until a dreadful fire wiped it out in 1892. The Sproul Lumber and Grist Mill flourished for many decades beginning in 1800; in 1873 the Mill was rebuilt and was owned by Mr. B.B. Haskell from 1884 to 1892. Next door on that same east side a few hundred yards above the Main Street bridge stood James Boyd's Iron and Brass Foundry which produced windlass parts, steering wheels, rudder hardware, spikes, bolts, and composition marine castings of all kinds from 1852 to 1892.

Finally, very near Main Street, north east of the bridge was a structure that stood through the 1950's and symbolically connected the town with the golden age of shipbuilding and manufacturing generally. This was the Stephen Jones Sail Loft where men fashioned the giant sails for local and out of town vessels. By 1927 the map indicates the terms "Tents and Awnings" had been added to its continued production of sails. (Today the Jones Sail Loft rests at the Maine Maritime Museum in Bath where it stores exhibits pertaining to the Lobster Industry. Thousands of visitors explore it every summer.)

Sproul's Mill in 1874; Jefferson Street Reed Mansion in background –
Lincoln County Cultural and Historical Society/Maine State Museum

The Lincoln County News – May 8 , 1879 – "Wood Carding! The carding machine at the Sproul Mill are now put in the most perfect order for carding wool. Farmers of Waldoboro and vicinity are informed that they can rely on having their work done in the best possible manner and at much reduced prices from former years by B.B. Haskell."

Medomak Flour Mill – Author's Collection

The Lincoln County News – November 19, 1880 – "The Medomak Flour Mill runs day and night. Mr. Arthur Miller of Augusta is the night miller."

Soule Mill (foreground); Medomak Flour and Saw Mill (background)
Courtesy Waldoborough Historical Society

Twitchell-Chapman corn and peas canning company – 1973 Waldoboro Bicentennial Book

The Lincoln County News – July 20, 1899 – “40 workers at the Twitchell-Champlin Cannery began their season’s work canning peas. For the next three weeks it will be one of the busiest places in Waldoboro.”

Soule Mill – Courtesy Waldoborough Historical Society

The Lincoln County News – July 26, 1877 – "Mr. Charles Soule has commenced removing the dam at his mill in preparation for building a new one. He produces carded wool, oakum, and ships' plugs."

The Lincoln County News – July 1882 – "A fire, starting from some oily rags on the second floor destroyed Soule's Carding Mill. There was no insurance, but Mr. Soule vows to have the mill running again soon." (By the end of August the frames of the new mill had been raised.)

Seven Merchants Of the Village

Mrs. Elizabeth Sproul of North Waldoboro recently loaned us a pamphlet which describes the towns of coastal Maine in 1892 historically as well as providing thumbnail sketches of the more prominent businesses and their owners. Our thanks to Mrs. Sproul for a springboard into another article about Waldoboro Village. (Note – Mrs. Sproul's grandfather, William Walter, was a noted businessman in his own right – a lumberman in North Waldoboro. He was also an amateur poet who once published his own book of poems.)
The seven Waldoboro Village establishments were praised by the writer of this pamphlet in the slow, rather ponderous style of the day, reminding us all of the much less hurried pace of life 100+ years ago.

On the steep east banking, a short way from the four corners of the village stood the M.M. Richards Co., "Jobbers of Foreign and Domestic Woolens, ready – made Clothing, and leading manufacturers of 12,000 to 15,000 pairs of pants and about 3,000 suits, ulsters, and overcoats per annum. Trade is widely distributed all over the state, and this concern is prepared to offer unsurpassed inducements to the most discriminating buyer." About 60 people were employed here – five at the factory, two on the road, and fifty women at home.

Above Richard's factory, on the street now known as Glidden – then Custom House Street – was J.T. Glidden's water powered Marble Monumental and Cemetery Stone Works. Mr. Glidden, whose son, E.A. was to continue the business after his father's death, was a native of Alna and connected with the stone cutting trade since 1855. His premises measured 135 feet by 40 feet and employed six people.

The enormous Fish Block stood on the corner of Jefferson and Main. George Bliss managed one of several stores in this block – one which specialized in books, stationery, paper hangings, borders, crockery, and fancy goods. Mr. Bliss, one of the most respected men in town, having served as collector of customs, and chairman of the Board of Selectmen, also offered the citizenry a lending library which contained "all of the standard works and the best literature of the day."

In 1892 there were two millinery shops at the Four Corners on Friendship Road. One, in the Sproul Block, belonging to Mrs. Clara S. Gay, had been started in 1888 and "has always made it a point to secure the services of courteous and faithful attendants, which always makes the task of shopping easier." The other shop, directly across the street, was owned by Mrs. S.E. Genthner and advertised "fashionable millinery, ribbons, laces, hosiery, gloves, Hamburgs (hats), ladies' and children's underwear, silks, velvets, worsteds; stamping done to order, fancy goods, small wares, and art materials." Mrs. Genthner's shop had been established in 1847 by her mother, Mrs. S.H. Balch, who had run it until 1865.

Two shoe stores were located in 1892 Waldoboro – both in the Fish Block. One was owned by E.V. Philbrook who claimed it to be the oldest business in town. The other, managed by Austin W. York since 1888, like Philbrook advertised "Ladies', Gents', Children's boots, shoes, and rubbers – also the repairing of same with all orders promptly executed." Mr. York ended his advertisement with the thoughts that promises of all salesmen from time immemorial – "In conclusion, we will remark that while we do not indulge in laudation (praise) of any house in particular, it shall be within our province to state to the public in general and our readers in particular, that if they would have their interests highly conserved, the acquaintance of this house should be made."

In other words much more familiar to folks 100 years later – "COME AND SHOP HERE; YOU WILL NOT BE SORRY!!"

Clothing for everyone; shoes for everyone, crockery, books, gravestones – offered by seven of the many shops located in the Village 100 years ago – all within easy walking distance.

Waldoboro is well represented by grocers, for they number many in all directions. As in every other branch of the mercantile activities, however, there are some good grocery establishments and some poor ones. Our province in this work is to point out to our readers the more deserving establishments in every line, and we are pleased to say that the grocery conducted by Mr. J. T. Gay in this town is one which we can recommend very highly. This enterprise was inaugurated in 1867 by Messrs. Mathews & Gay, who carried it on till 1870, when the firm was dissolved and the present proprietor assumed control. He is a native of this town, and the premises occupied are two stores 30x60 feet and one large storehouse 30x40 feet in dimensions. Two competent assistants are employed, and a large wholesale and retail trade is done in choice family groceries, paints, oils, glass, brushes and cordage, stoveware, flour, feed and grain, salt by the carload, also tobacco and cigars, etc., of which a large stock is carried. Mr. Gay obtains his groceries from the most dependable sources, and as he is favored with a large and regular patronage the stock is continually in rotation, which needs replenishing frequently; this assures the freshness and purity of the goods at all times. Mr. Gay's prices are as low as any in town and his terms easy. He is one of our leading and most enterprising business men and deserves to be patronized.

C. S. SOULE, Wool Carding promptly attended to, Waldoboro, Me. The business conducted by Mr. C. S. Soule may be said to date back to the beginning of the century, for it was founded about the year 1800, but like nearly all enterprises of such great age it has changed so in accommodating itself to the changing conditions of the times as to bear but a remote resemblance to its original form. The present proprietor has had control since 1842, and it is no exaggeration to say that he is universally known thoroughout this section, for during the past half-century he has carded a large proportion of the wool produced in this vicinity, which business he still follows, although in 1887 he discontinued dressing cloth, after 45 years of experience in such work. He is prepared to do wool carding at very short notice, and during the harvest season runs a machine for threshing grain of all kinds. He also runs a machine that does family washing (but not the washing of families), with "neatness and despatch," and that cleans carpets thoroughly and without injury. Ample water power is available, and all work is done promptly, as well as at reasonable rates. All packages sent by rail or otherwise given prompt and personal attention.

FRANK ACHORN, Grist Mill, Waldoboro, Maine. It is unnecessary to dwell upon the importance of being able to purchase Grain, Flour, Feed, etc. in any desired quantities at the lowest market rates for the advantages derived from an enterprise which has for its object the furnishing of an abundant and dependable supply of those staple commodities at bottom prices are so obvious as to be understood by every member of the community. Therefore it goes without saying that the establishment conducted by Mr. Frank Achorn, is popular throughout this vicinity, for this is a finely equipped steam grist mill, and the proprietor is in a position to meet all honorable competition in his line of business, filling the most extensive orders at short notice and always quoting prices in accordance with the lowest market rates. The mill is a structure of the dimensions 36 x 34 feet and is fitted up with improved machinery driven by a thirty five horse power engine thus orders can be filled with equal facility at all times of year. It is conviently located, having side track to mill and employment is given to one efficient assistant. Mr. Achorn, is a native of this town and has been connected with his present enterprise since 1890.

BUILDERS ATTENTION. W. R. WALTER, Lumber Manufacturer, Pine, Spruce, Hemlock, etc. All dimensions to order. No. Waldoboro, Me. One branch of industry creates or keeps another, and they all tend to the prosperity of the town and the good of the country. Builders and carpenters use lumber for their houses and for this reason the saw mill has to be brought into use. Prominent among those engaged in is line of business in North Waldoboro is Mr. W. R. Walter who has been carrying it on since 1889. The mill owned by this gentleman is located in a favorable position and driven by water power, and Mr. Walter is prepared to fill orders for lumber of all dimensions on reasonable terms, Frames sawed to order, Pine, Spruce and Hemlock boards and Shingles constantly on hand. Job planing and custom sawing is also done at short notice and moderate charges. Mr. Walter is a native of Waldoboro and during the time he has been in business here has made many friends and all he asks is to give him a trial and he will try to give satisfaction. Mr. Walter through the harvest season runs a threshing machine at his mill which draws trade from the surrounding country eight to ten miles.

EVERETT TROWBRIDGE, Dealer in Groceries, Crockery and Glassware, Fruit and Confectionery, Waldoboro, Me. The enterprise conducted by Mr. Everett Trowbridge must be given a leading position among Waldoboro's representative mercantile undertakings, although inaugurated only a few years ago yet has been so ably and honestly managed from the first as to have gained and held the entire confidence of the public. The premises occupied are centrally located in this town, and comprise one floor 50x22 feet in dimensions, the space available being none too large to accommodate the stock carried gives some idea of its magnitude, but to appreciate its variety and completeness personal inspection is indispensable, for it comprises a full line of general merchandise, including groceries, fruit and confectionery, crockery and glassware, etc. People buy here because the assortment is so large all tastes can be suited, because the employment of two assistants assures prompt and polite attention to all, because the prices are uniformly moderate, and last, but by no means least, because the goods are sold strictly on their merits and will prove just as represented. Mr. Trowbridge is a na- of Damariscotta, Maine.

GARDNER J. NASH, A General Assortment Harness Goods, Waldoboro, Me. It is both safer and cheaper to use a good harness than a poor one, and there need be no difficulty about getting a good harness if you go to the right kind of a place. We have no hesitation in recommending that now carried on by Mr. Gardner J. Nash, in this town, for this establishment

was founded by him in 1888. Mr. Nash is both able and willing to produce fine harness at prices as low as the lowest for similar work. His establishment is one floor 40x20 feet in dimensions, and is completely fitted up for the manufacture of harness of every description, and the doing of repairing in general. A well selected stock is constantly carried, comprising all kinds of harness goods and horse equipments, etc., these articles being offered at moderate prices as well as guaranteed to prove as represented. Mr. Nash is a native of Nobleboro, Me., and has a large circle of friends throughout the vicinity, and numbers among his customers some who have traded with him for many years. His reputation for turning out

RIVERSIDE HOTEL, Chas. A. Jackson, Proprietor, Jefferson Street, Waldoboro, Me. The Riverside Hotel holds a high position among hotels which offer home like accommodations to the travelling public, and indeed it would be difficult to find a house in which patrons are actually given "more for their money" in every sense of the phrase. Of course Maine has many first-class hotels, furnished up in grand style, equipped with "all modern conveniences," etc., but most of us don't care to live in palaces, and are perfectly satisfied when we can get a pleasant, well kept room in a well managed hotel, where the table is supplied with an abundance of well cooked and neatly served food. Now this is just what is obtained at the Riverside Hotel, and therefore we take pleasure in recommending this hotel to those who appreciate "solid comfort" and wish to avoid exorbitant prices. The hotel in question was originally opened in 1879, by Mr. Tilson Benner, as the "Benner House," and after several changes, in 1891, passed under the management of Mr. Charles A. Jackson, the present proprietor, giving the hotel its present name. Mr. Jackson is a native of Jefferson, and has a thorough knowledge of the art of hotel keeping. The house, which is a three story brick building, and basement 50x40 feet in dimensions is well furnished and newly painted, papered, etc., has a parlor, dining room, and nineteen light and airy sleeping rooms. Ten competent and courteous assistants are constantly employed, and every attention is paid to the wants of guests, either permanent or transient. A first-class livery stable is connected which has been run by Mr. Jackson for two years, giving room for twenty horses, which may be hired at reasonable rates.

Representative Business Men of Waldoboro Advertisements from a 1892 pamphlet – Courtesy of Mrs. Elizabeth Sproul

J. P. GLIDDEN, Water-power Granite and Marble Monumental Works, Waldoboro, Maine. Designs and Estimates furnished. If there is one branch of industry in which special attention should be paid in avoiding any appearance of "cheapness" in the finished product, it is that of monumental work, for nothing is in worse taste than an elaborately designed monument or headstone, the mechanical work of which is done in a rough and careless manner, and on the other hand an inappropriate design will produce a bad effect, no matter how carefully and skillfully it has been carried out. The importance of placing orders with a person who combines tasteful designing with excellent execution is therefore obvious, and it is not at all surprising that Mr. J. P. Glidden should be liberally patronized by the residents of Waldoboro and vicinity, for this gentleman turns out work equal to the best in every respect, and quotes the very lowest prices. Mr. Glidden is a native of Alna, Maine, and has been identified with his present establishment since 1855. He is well known and highly respected throughout the community. Mr. Glidden carries on granite and monumental marble works, and using the best qualities of American and Foreign kinds. The premises utilized are large, being 135x40 feet in dimensions, and having water power to drive his machinery. Designs and estimates are cheerfully given, and out of town work is solicited. Six assistants are employed.

W. H. LEVENSALER & CO.,

MANUFACTURERS OF

CUSTOM and READY-MADE CLOTHING,

AND DEALERS IN

Gent's Furnishing Goods, Hats, Caps, Trunks and Paper Hangings.

HIGHEST PRICES PAID FOR WOOL.

WALDOBORO, MAINE.

There is a real art in knowing what the people want, and when this knowledge is had and intelligent measures are taken to supply that want it of course follows that a popular enterprise is built up. No one to see the amount of business done by W. H. Levensaler & Co. would think that this undertaking had only been under their management since 1880, but such is the fact, and the extensive wholesale and retail trade now carried on speaks volumes for the tact and ability of the proprietors who are natives of this town and well known, and at their establishment located in Otis Miller's block they are prepared to offer a fine assortment of custom and ready-made clothing, furnishing goods and gent's underwear at very low prices. They also manufacture indigo blue shirts, pants and felt goods in frocks, vests and shirts, and announce that they have all the variety of styles and qualities necessary for the comfort and vanity of man, and an inspection of their stock will show that the announcement is no idle boast. It will cost no one a cent to compare Levensaler & Co.'s goods and prices with those offered at other establishments. The premises used are two floors 55x25 feet and ten competent assistants are employed.

MISS CLARA S. GAY, Millinery and Fancy Goods, Waldoboro, Me. Anyone at all familiar with the stores in this community knows where to locate Miss Clara S. Gay's place of business. This attractive establishment was founded by her in 1888. Her enterprise found favor with the public from its first inception and increased from year to year, until now she has little available space left. to accomodate the large stock carried she now occupies one floor 35 x 22 feet in dimensions and has always made it a point to secure the services of courteous and faithful attendants and has retained their services at all times. The sight of familiar and pleasant faces behind the counters of an establishment of this kind renders the task of shopping more agreeable. The stock carried by Miss Gay embraces a full and varied assortment of high class Millinery and Fancy Goods which cannot be surpassed if indeed equalled either in quality or price in this town. Miss Gay is an estimable lady and a native of this town and her enterprise in our midst should meet our approval in liberal patronage.

Balch's Corner – Author's Collection

The Lincoln County News – July 28, 1882 – "Speckaree, the Italian peanut vendor, has, because of the unusual heat this summer, moved from across the street at Balch's Corner to the corner of Main and Jefferson where his nice fruits and peanuts are always welcomed."

The Lincoln County News – March 16, 1899 – "The disturbance on Austin's Corner for an hour Monday afternoon was no credit to the town."

The Fish or Union Block – largest block in Lincoln County
1973 Waldoboro Bicentennial Book

Clint Stahl (right), hotel proprietor, actor, store manager, musician, baseball player, sits with friends at the barber shop. Courtesy Waldoborough Historical Society

The Lincoln County News – October 19, 1899 – "Clint Stahl is now selling crockery, glassware, lamps, and novelties in his shop in the Sproul Block."

M. M. RICHARDS & CO., Jobbers of Foreign and Domestic Woolens, Ready-Made Clothing, Waldoboro, Me. Nothing impresses observing visitors to this country more than the well dressed condition of the people, for it is not paralleled in any other country in the world, it being a conceded fact that Americans are the best dressed people on earth. The reason generally given for this state of affairs is the great prosperity of the nation, and, of course, that has a great deal to do with it, but there is another important reason,—the cheapness and excellence of ready-made clothing in this country. American ready-made clothing is the best fitting, the best made, and, quality considered, the lowest priced in the world, and this assertion can be proved. despite the claims of those who assert that better and cheaper clothing is produced abroad. In this connection we may fittingly make mention of the firm of M. M. Richards, & Co., for they are leading manufacturers and jobbers of ready-made clothing, and no concern in the state is prepared to furnish equally desirable clothing at lower prices than they quote. That their facilities are appreciated and availed of is partially indicated by the fact that they manufacture from 12,000 to 15,000 pairs of pants, and about 3,000 suits, ulsters and overcoats per annum, and we say partially indicated, as this is but one department of their business, they being extensive jobbers of foreign and domestic woolens; they are the sole agents of the celebrated Knox Woolen Co.'s gray and indigo blue satinets. Their trade is very widely distributed, and not only its magnitude but its character prove that this representative concern are prepared to offer unsurpassed inducements to the most discriminating buyers. Mr. Richards is a native of Temple, Me., has had long experience in his present line of business, and is favorably known to the trade throughout this portion of New England.

Waldoboro Pants Factory – M.M. Richards – Opened in 1893
1973 Waldoboro Bicentennial Book

The Lincoln County News – January 4, 1900 – "M.M. Richards is preparing sample business cards. This is one of our flourishing industries; wish we had more of them."

L.C.N. – February 15, 1900 – "M.M. Richards is shut down for a week or so."

Sproul Block – 1916 – Author's Collection

"Traveling About" (Part 1)

Faster, dependable, and more comfortable means of transportation has forever been a goal of human beings. A speedier horse, dog team, and sailing ship preceded later inventions – steam ship, railroad, automobile, and airplane whose modern day speeds and passenger comfort would be totally inconceivable to our ancestors.

Waldoboro folks of the late 1800's of course did not have access to all this speed and comfort that we take for granted. Understanding that, it has been interesting to study their modes of transportation through the reports in *The Lincoln County News*, the village paper of the day. What they did have was a surprising number of choices – inexpensive if slower and more time consuming. More peaceful perhaps? Judge for yourself!

From the files of the *L.C.N.* we learn that

"A party young men came around Thursday from Thomaston to Waldoboro in the steamer *Ocean*. They put up at the Medomak House (Editor's note – The Medomak House was where the Grange Hall parking lot is today on Main Street) and returned home Friday." 7/31/1879

"Sleighing parties are now able to "coast" from the top of Cole's Hill to and across the Bridge on Main Street." 1/18/1900

"Excursion train – Waldoboro – Castine – to hear a concert given by the Thomaston band. $1.00" 7/23/1878

"WALKING to vote at the Town House several miles were Henry Kennedy (ae 75), Alfred Hovey (ae 85), George Smouse (ae 75), and Fred Castner (ae 94)." 9/18/1874

"Moses Levensaler WALKED to Ezra Genthner's house and stayed over for the night – then walked back – a total of 18 miles." 9/27/1877

"A 16 year old boy WALKED from Rockland to Friendship then to Waldoboro and thence to Damariscotta. His father found him there in a selectman's home seemingly not exhausted from the 35 mile trip." 3/8/1900

"A trolley line now runs from Camden to Rockland to Thomaston to Warren. We understand that it will eventually run clear to Boston." 3/13/1902

"Dust is now 6" thick on Waldoboro streets. We need a street sprinkler." (Editor's note – The sprinkler idea was voted down!) 5/29/1902

"The new steamer *Bristol* will leave Storer's Wharf at 8 A.M. for Monhegan. $1.00 per ticket." 8/8/1901

"Street signs and gas lamps can be seen for the first time on Jefferson, Union, Depot, and Elm streets." 6/16/1882

"A special train will run to Rockland this Friday. 'HMS Pinafore' will be performed. Train leaves at 5:10 and returns after the performance. $1.00 for ride and play." 4/3/1879

"Lumber for town sidewalks has arrived finally. Are sidewalks for pedestrians or for bicycles? A lady and a child were nearly injured last year. It is not a pleasant sensation to have riders approaching at a top rate of speed and think you may be run down. It is not allowed in other places. Why here?" 6/6/1901

"Stages leave Waldoboro from the Medomak House at 11:30 A.M. North destinations are Jefferson, Cooper's Mills, North Whitefield, South Windsor, and Augusta. These coaches are also mail carriers; they are driven by gentlemen and are very comfortable."

"Some regularly scheduled Maine Central Railroad trains through Waldoboro Depot: 7:40 A.M. Rockland; 8:26 Waldoboro for Bath, Brunswick, Lewiston, Augusta, Waterville, Portland, Boston. 4:00 P.M Parlor Car for Boston
1:30 PM Rockland – Waldoboro to arrive in Boston at 9:20.
Some regularly scheduled Boston and Bangor Steamship Line boats: Rockland – Boston – Monday and Friday at 6 P.M.; Tuesday, Thursday, Saturday at 7:30 P.M. Returns from Boston every week day except Thursday at 5 P.M.
Also Rockland – Camden, Belfast, Bucksport, Winterport, Bangor – leaves Wednesday – Sunday 6 A.M.; Tuesday, Thursday, Saturday 5:00 A.M." 6/6/1896

Waldoboro, Maine – Main Street – Photo taken in 1876 – Medomak House with stage
Author's Collection

The Lincoln County News – August 7, 1874 – "Stages leave Waldoboro every day at 11:30 A.M. for Jefferson, Cooper's Mills, North Whitefield, South Windsor, and Augusta. Comfortable coaches, gentlemanly drivers; also carries mail."

Traveling About (Part 2)

Three accounts of excursions taken by citizens of Waldoboro in the late 19th century give us a feeling for the time, effort, and sense of adventure such trips would entail. Readers, please think how the arrival of the automobile with its speed and sense of independence would change these citizens' lives in the years shortly to come.

In late June of 1878 a group of Waldoboro families set forth on a trip to Boothbay Harbor by horse and buggy. They spent the night before in Damariscotta at friends' houses and then, arising at 6:30 A.M they struck out down the River Road. Reaching the Harbor at 10:00 A.M., they spent time shopping, bowling, and playing some billiards. At 11:00 A.M. they boarded an excursion boat for Mouse and Squirrel Islands where they attended a band concert and a dance. Returning to Boothbay after these island festivities around 8:00 P.M. they climbed aboard their buggies, retraced their journey on the dark River Road, and arrived at their hosts' homes at midnight. The next day they returned to Waldoboro.

A second excursion embarked from Storer's Wharf, Waldoboro, in July 1877. Captain E.O. Winchenbach's yacht *Bee's Wing* had been chartered by an intrepid group to visit some of the islands in Muscongus Bay – Long, Friendship, Cranberry, and Harbor among others. A sumptuous bean picnic was planned, and all supplies for this feast were aboard as the jolly group tacked down river. Somewhere off Dutch Neck misfortune struck; a heavy squall capsized *Bee's Wing*, and crew as well as captain, floundering in the cold water, had to be rescued by an alert band from the mainland. The picnickers were saved – wet but alive. Sadly, the entire bean lunch, baskets, tablecloth and silverware were not recovered.

In April of 1899 it was announced by *The Lincoln County News* that Waldoboro folks had an opportunity to travel by train and boat to Boothbay. There several local high school oratorical teams including Waldoboro were participating in a contest. For a fee of $1.25 interested citizens could board a train at Waldoboro, and journey to Woolwich. There, as part of the package, they would steam by boat up to Boothbay, enjoy the speaking contest, and then return by boat and train to Waldoboro.

Those who made the trip were rewarded – Waldoboro won the contest! When the students' train arrived home that evening, they found that the news of their victory had preceded them, and that an impromptu celebration was under way. The contestants were met at the Depot by a fleet of carriages full of proud townspeople and decorated with flags and bunting. As the parade wound up Jefferson Street and through the Village, chiming bells from the churches and the High School mingled with steam whistles from the shoe factory and M.M. Richards pants factory. At the Baptist Church a reception for the champions was held, brief congratulatory remarks expressed, and light refreshments served. A wonderful spur-of-the moment event in Waldoboro history!

Waldoboro Railroad Depot – "The students' train was met by a parade of vehicles."
Courtesy Waldoborough Historical Society

The End of an Era – Horses to Automobiles

The significance of the following excerpt from *The Lincoln County News* – January 28, 1904 is worth pondering:
"We understand that Waldoboro will have one or two automobiles in town by next summer. It would be well to get your horses educated to them at the first opportunity." Little was anyone – in Waldoboro or elsewhere – aware of the unbelievable changes to come after thousands of years of dependence upon one animal.

When families came to town in 1884, they arrived as they always had – in their horse driven vehicles. If they were remaining for a brief period, they would park at curbside. However, most people planned on spending the day in the village and needed a place to keep the horse and buggy while they walked, shopped and visited. Such a place was called a livery stable, and there were several available. The Exchange Hotel, part of the Union or Fish Block, corner of Main and Jefferson had a stable across the street. The Medomak House Hotel had one in the present Grange Hall parking lot. Benner's Stable stood on Friendship Road.

Stan Hanna, long time resident, remembers Benner's Stable: "There was an annex on the right hand side where the office was; vehicles could be parked inside the building on either the right or the left. Way out back were the stables for the horses." Bill Blodgett, another fairly long term resident, had the task of tearing down Benner's Stable in the 1960's. He remarks about the solid construction and the many empty bottles he found hidden in the walls. (One editorial in the *LCN* complains about the rowdiness in terms of drunkenness and fistfights that were occurring at the stable on occasion!)

Horses (and oxen) had to wear iron shoes to protect their hooves. Shoeing these animals as well as producing other metal implements were the jobs of blacksmiths. There were many of these clever men in Waldoboro. Stan Hanna remembers the sons of William White – John, Lewis, and especially – Ephraim, the most skilled, said to have fashioned a magnificent pistol when he was but 11 years of age and who was hired by the U.S. government to make tempered steel bits used in the building of the Panama Canal. Ephraim's shop was found at the east end of Elm Street, corner of Depot. Lewis and his father occupied one half of the building used now by the "Bad Hair Days" beauty salon. Another blacksmith was Landon Waltz who used the present Eugley home next door to True Value Hardware. There were shops in North Waldoboro (Lexington Mank), Route 220 (John Shuman), Route 32 to West Waldoboro (Benjamin French), West Waldoboro (G. Burton Hilton), and South Waldoboro (Oliver Studley). For the time being – plenty of work for all.

By the 1920's, the long era of man's total dependence on the horse was coming to a close (although not by any means finished, according to Stan and others who experienced it). Street maps of Waldoboro village show clearly some of the changes being brought about by the ever improving automobile. In 1927 two filling stations are shown – one on Pleasant Street across from the Baptist Church and the other on Jefferson Street just across from "Bad Hair Days". Two parking/repair garages have also appeared – one for 70 cars on Jefferson Street in (you guessed it) "Bad Hair Days Salon" and the other – for 40 cars – where Maine Antique Digest is found today.

Who would have conceived of such changes in 1903 – 100 years ago?

Ephraim White in his blacksmith shop
1973 Waldoborough Bicentennial Book

Two Disastrous Fires

Waldoboro's history, like so many other towns, is marred by incidents of fire. In 1846 nearly the entire business section was destroyed; in 1854, in an even worse calamity, almost every store and workshop, and even ships on the stocks were leveled, and 70 families were made homeless. In both situations energy and money came to the rescue – the shipbuilding boom was at its peak – and the village was rebuilt almost immediately.

By 1890 Waldoboro shone in Lincoln County as a bustling industrial place boasting of: a shoe factory, an iron and brass foundry, a flour mill, several lumber mills, an oakum mill, a granite quarry, a corn packing plant, a shirt factory, a pants factory, a steam laundry, a carriage and paint shop, a steamship business, a suspender shop, and several blacksmiths. The streets and sidewalks teemed with citizens, visitors, dogs, cats, cows, and horses; one had to stand in line for even an ice cream cone – ESPECIALLY on Saturdays when the farmers came to town for supplies.

This busy picture was to remain for many years; however, the town landscape changed for good in the 1890's as a result of fire. This time there were no wealthy ship builders to rebuild as they had done in earlier years.

The first fire broke out at 11:30 P.M. April 23, 1892 at the "Lower Falls" just above the Main Street bridge. Here sat two mills – the grist, saw, planning and plaster mill of McIntyre & Son (William Sproul had begun the business 100 years before) and James P. Boyd's Iron and Brass Foundry (established in 1855 and manufacturers of spikes, bolts, steering mechanisms and general castings). The rapidly spreading flames quickly consumed these structures along with William White's and Ephraim Simmons' blacksmith shop and adjoining brick residence. Neither McIntyrc nor Boyd carried any insurance.

A littlc more than one year later came the second and even worse fire. At 11:00 P.M. on the 12th of June, 1893, black smoke was seen coming from the roof of a small stable south and west of Jefferson Street. Soon J.K. Willett's furniture storerooms, his store, Orrin Achorn's tenement house, and the stables of the 30 room Medomak House on Main Street were ablaze. Inexorably the conflagration spread to the hotel itself; the whole area north of lower Main Street and west of Jefferson Street was being gutted. The little "Triumph" water pump was placed in the Baptist Church reservoir, and volunteers ran the hose all the way to the Medomak House. Unfortunately, after a half hour the machine gave out; it seemed as if once again the entire village was to be at the fire's mercy. Somehow the citizenry managed to confine the blaze to the Jefferson Street line.

Eleven families lost their homes in this fire with the total property loss estimated at $33,500. The Medomak House, stagecoach and mail stop since 1855, was gone. Smoke drifted up from the ruins for weeks with many out of town visitors arriving to "gawk".

Eventually the town built a parking lot in the hotel's site and then a Grange Hall. Of the two riverside mills, McIntyre's and Boyd's, nothing today remains except the scattered rocks of the old foundations.

Material for this article was procured from J.J. Stahl's *History of Old Broad Bay and Waldoboro*, S. Miller's *History of Waldoboro*, the pamphlet "Waldoboro", the Waldoboro Bicentennial Book, and *The Lincoln County News*.

Ruins of the McIntyre (Sproul) Mill area after the 1892 fire. 1973 Waldoboro Bicentennial Book

FIRE IN THE FISH BLOCK!!!!!

The tragic loss of the Down Easter Inn in Damariscotta and the subsequent comprehensive writeups stirred this author to investigate similar happenings in Waldoboro. In doing so it was interesting to compare modern fire fighting techniques with those of the past – and to be grateful for what we have. (Files of *The Lincoln County News* and Jasper Stahl's *History of Waldoboro* are the sources for the following story.)

June 1900 – On the southwest corner of Depot (Jefferson) Street and Main Street and across from Clark's Hall stood the largest business block in Lincoln County. This was the Fish Block (also called the Union or the Matthews Block) – a substantial brick and granite structure – three stories on Depot Street, four on Main Street, and five in the rear. It had been built in 1855 by William Fish and John W. Tibbets, owned by Mr. Fish until his death in 1880 and then by Moses R. Mathews of Warren. Businesses in the block in 1900 included The Exchange Hotel, the Knights of Pythias Hall, an American Express Telegraph Agency, the Medomak National Bank, N.C. Austin's large department store, a grocery owned by J.S. Overlock, and George Bliss's department store.

Mr. Stahl's remembrances of the terrible morning of June 5, 1900 follow: "It was a bright and pleasant morning. As I was making myself ready for school, the church bells in the village started their noisy clangor (the signal for a fire). The sound carried far up and down the valley in the clear and quiet morning, and soon the suburban population was pouring hurriedly into the village to man the pumps on the old tubs. The fire had started in the attic of the Union Block, and shut off as it was from air, it burned slowly downward, and burn it did, and by noon the south and east walls had collapsed into the streets."

Because it was soon evident that the two small town engines, Triumph and Medomak, hooked up to reservoirs by the Baptist Church and on the Haskell property, would be unable to quell the blaze, telegraphed messages were sent to Bath, Rockland, and Thomaston for assistance. The message to Rockland was received at 8:30, and a special train left at 9:19. Picking up Thomaston's Eureka engine to go with their own, the Rockland men arrived in Waldoboro's depot (still quite a distance from the fire) at 9:42. Eventually placed at the river's bank, this equipment contributed largely to the saving of buildings across Main Street and down from the Fish Block. Bath's effort to send help were unfortunately negated by the facts 1) Maine Central had not received any orders for a special fire fighting train, and so none was ready at the Woolwich depot. 2) Low tide prevented another locomotive to be brought over from Bath by steamboat (Remember – there was no bridge available in 1900!). So Bath firefighters, ready to embark, were unable to. Finally, Waldoboro's telegraph office being inside the burning building made it impossible to receive or send out messages.

The fire worked down from story to story. Whenever a floor fell, showers of sparks were taken by the southwest wind and carried to other buildings which caught fire but were promptly extinguished. The north wall was the first to fall, followed by a portion of the east wall, and finally the south wall, four stories high, which fell into Main Street with a terrible crash. Miraculously, none of the many spectators were hurt. Much debris – bricks, slates, and burning embers littered Main Street hill.

Following are figures of losses to the merchants of Fish Block and their compensary insurance coverage:

Name	Loss	Insurance coverage
M.R. Matthews	$20,000	$8000
N.C. Austin	12,000	5200
George Bliss	7500	4000
Maine Lodge	500	1000
American Exchange Telegraph	small	100
J.S. Overlock	1500	800

The Fish Block was never rebuilt to match its former grandeur. Mr. Mathews, after a two year hesitation, had a large wooden structure constructed and named it after his partner in the project, Mr. John Gay. This building lasted until 1960, when it too burned.

If they had been available to use, modern fire extinguishers would without question have taken early control of the slowly burning attic fire discovered early in that historic day, June 5, 1900.

Ruins of the Fish Block – 1900 – Courtesy Waldoborough Historical Society

Corner of Main and Jefferson Street – Gay Grocery in the foreground (where Stan Bailey ran his roasted peanut business). The other white building (Taction) was the clothing factory. Author's Collection

Looking up Main Street toward the Customs House and Reed Mansion.
Author's Collection

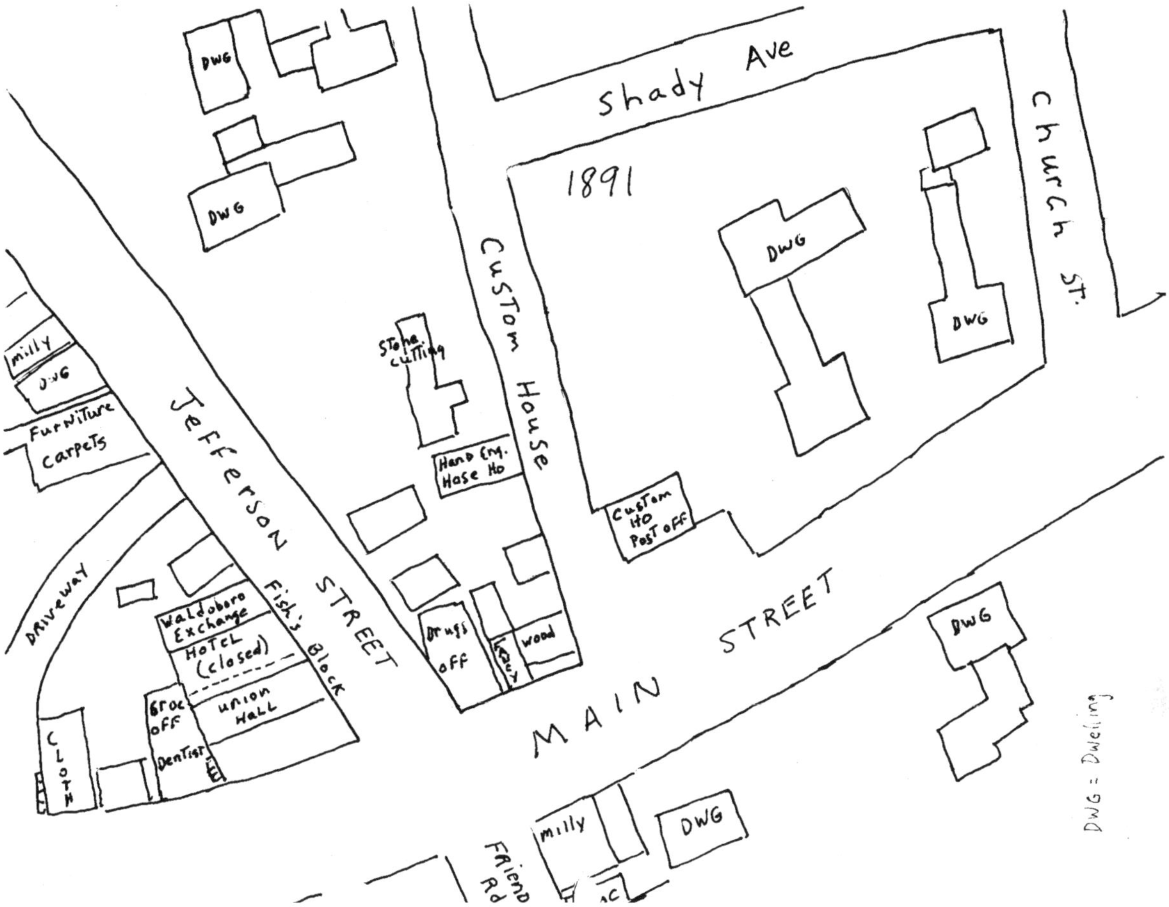

The Village changes…1891–1927
(A series of maps, pp. 211-214; 216-221)

Note the losses of the Medomak House, the Fish Block, Sproul's Mill area, and livery stables… caused by the fires of 1892, 1893, and 1900. Maps all courtesy of Waldoborough Historical Society.

Note also the gains – Odd Fellows Block, Gay Block, a movie theater (to replace Clark's Hall), and garages for the new fangled automobile!

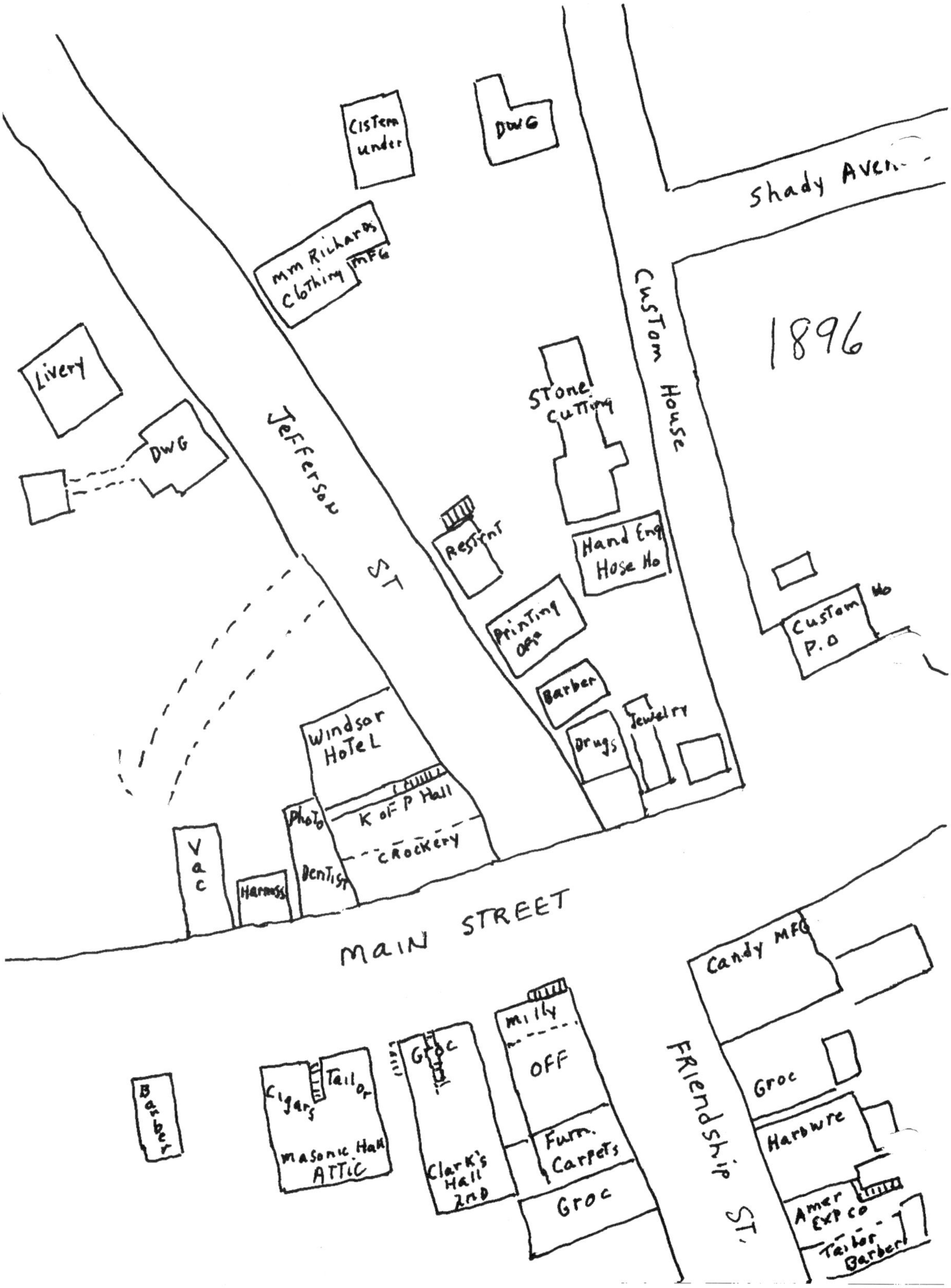

Cistern under
DWG
Shady Aven
Mm Richards Clothing MFG
Custom House
1896
Livery
DWG
Stone Cutting
Jefferson St
Restnt
Hand Eng Hose Ho
Printing Off
Custom P.O
Ho
Barber
Jewelry
Drugs
Windsor Hotel
K of P Hall
Photo
Crockery
Vac
Harness
Dentist
Main Street
Candy MFG
Milly
OFF
Groc
Tailor
Cigars
Barber
Masonic Hall ATTIC
Clark's Hall 2nd
Furn. Carpets
Groc
Friendship St.
Groc
Hardwre
Amer Exp Co
Tailor Barber

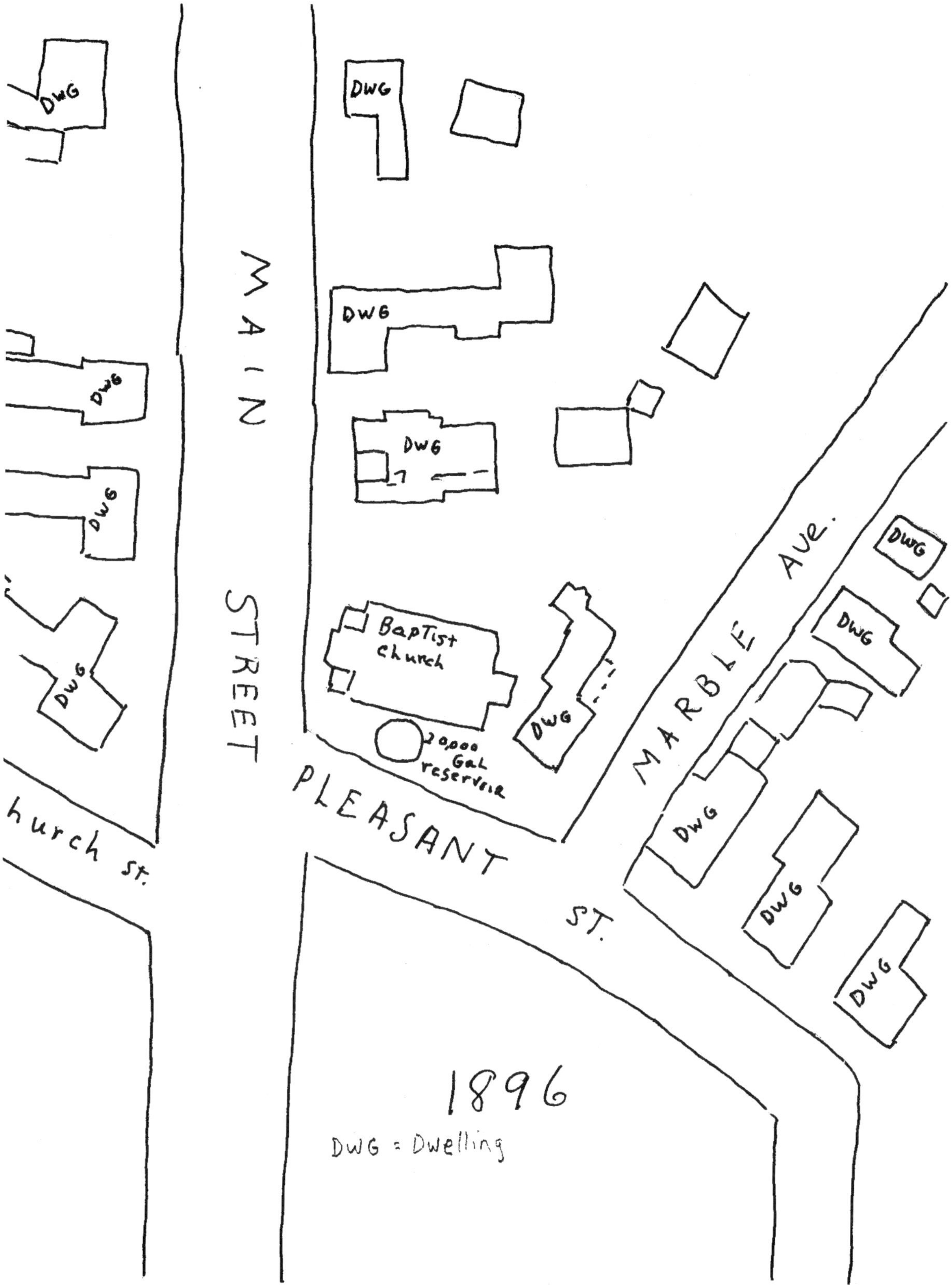

DWG
DWG
MAIN
DWG
DWG
DWG
DWG
STREET
BapTist church
DWG
DWG
MARBLE AVE.
DWG
DWG
20,000 GaL reservoir
DWG
DWG
DWG
DWG
hurch St.
PLEASANT ST.
1896
DWG = Dwelling

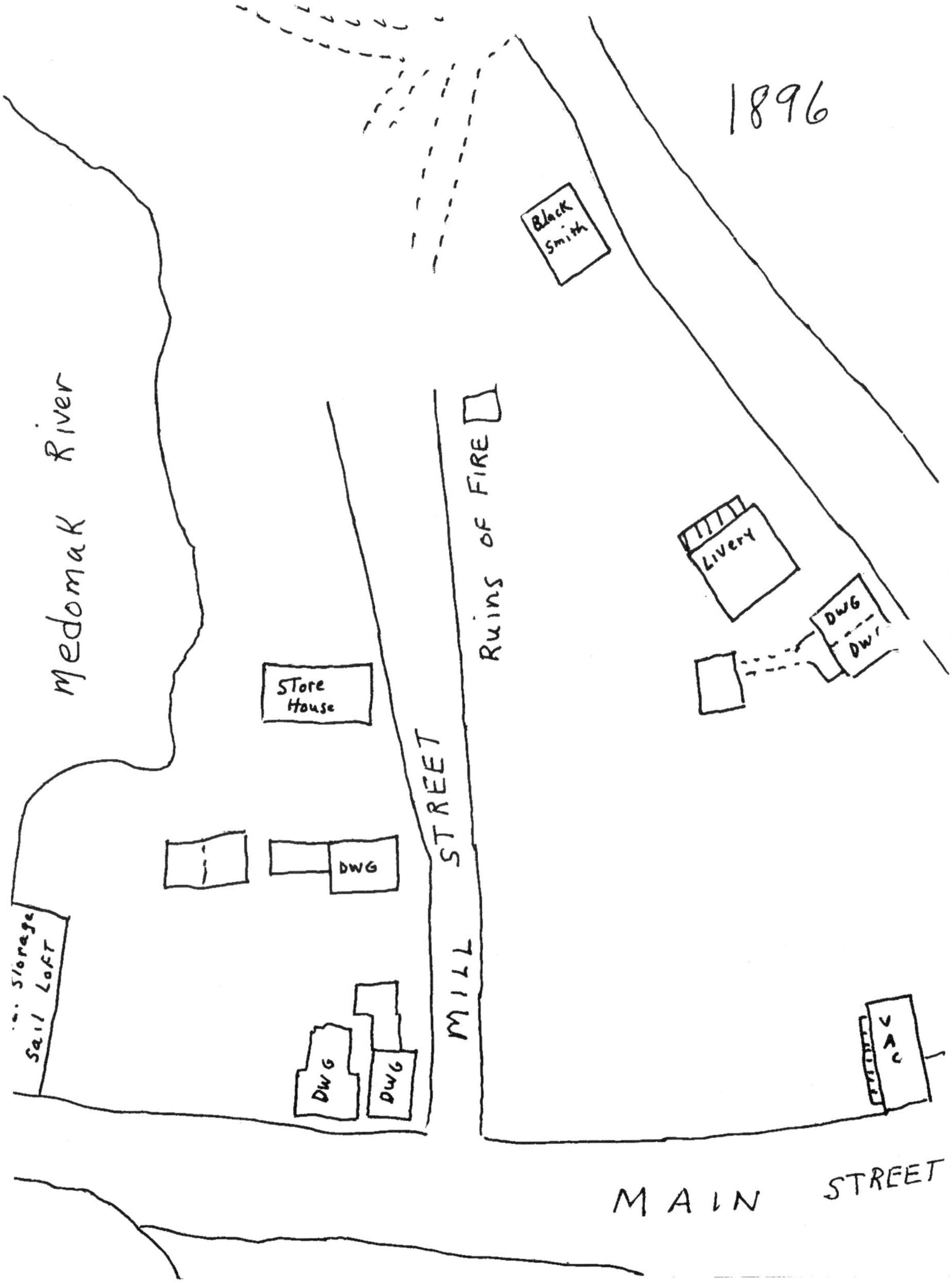
1896
Medomak River
Black Smith
Ruins of Fire
Livery
DWG
DWG
Store House
MILL STREET
DWG
Sail Loft
DWG
DWG
MAIN STREET

The Shoe Factory – grand hope to replace shipbuilding – failed – eventually became the Button Factory. – Author's Collection

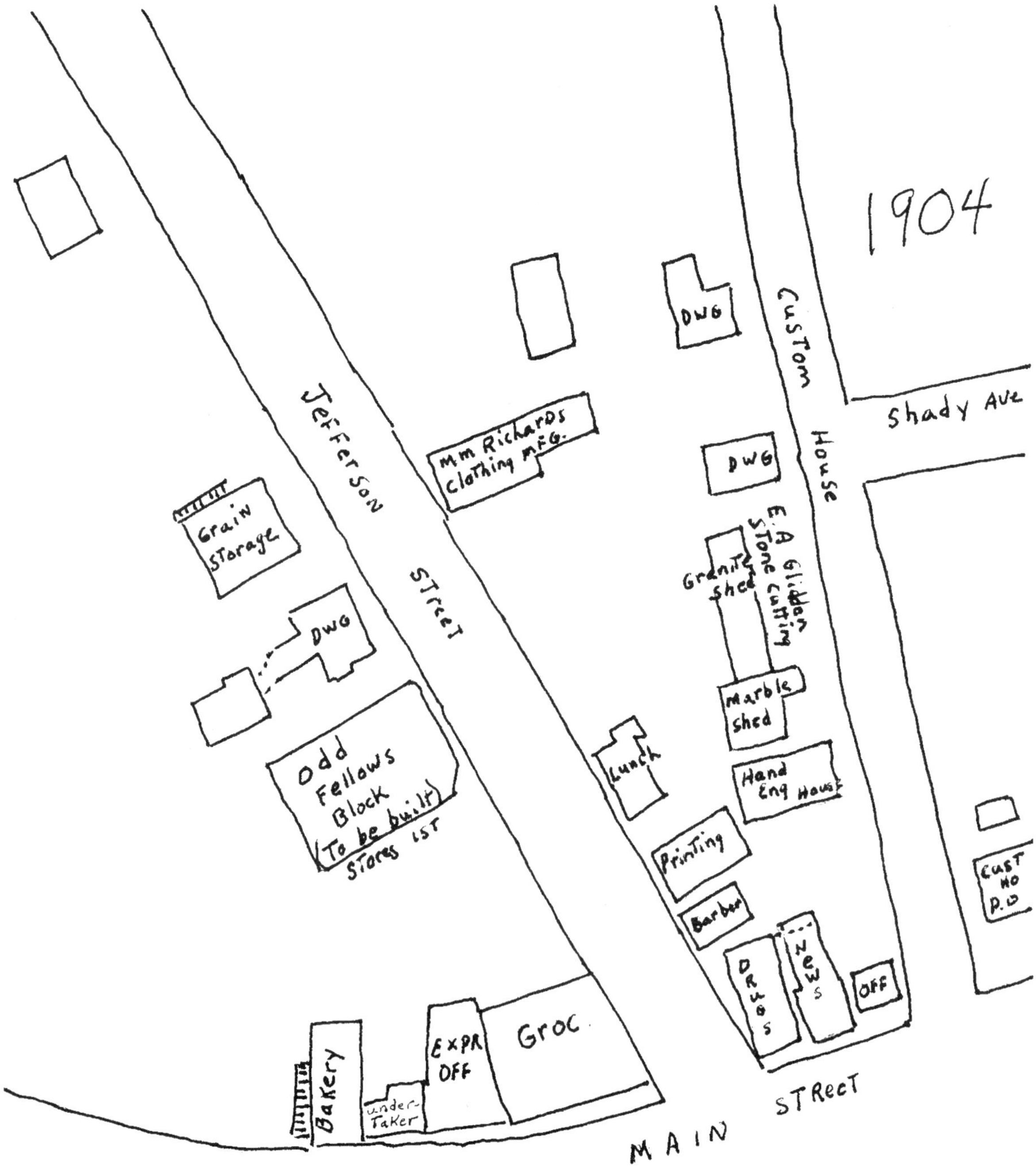

Gay's Grocery replaces the Fish Block which burned in 1900.

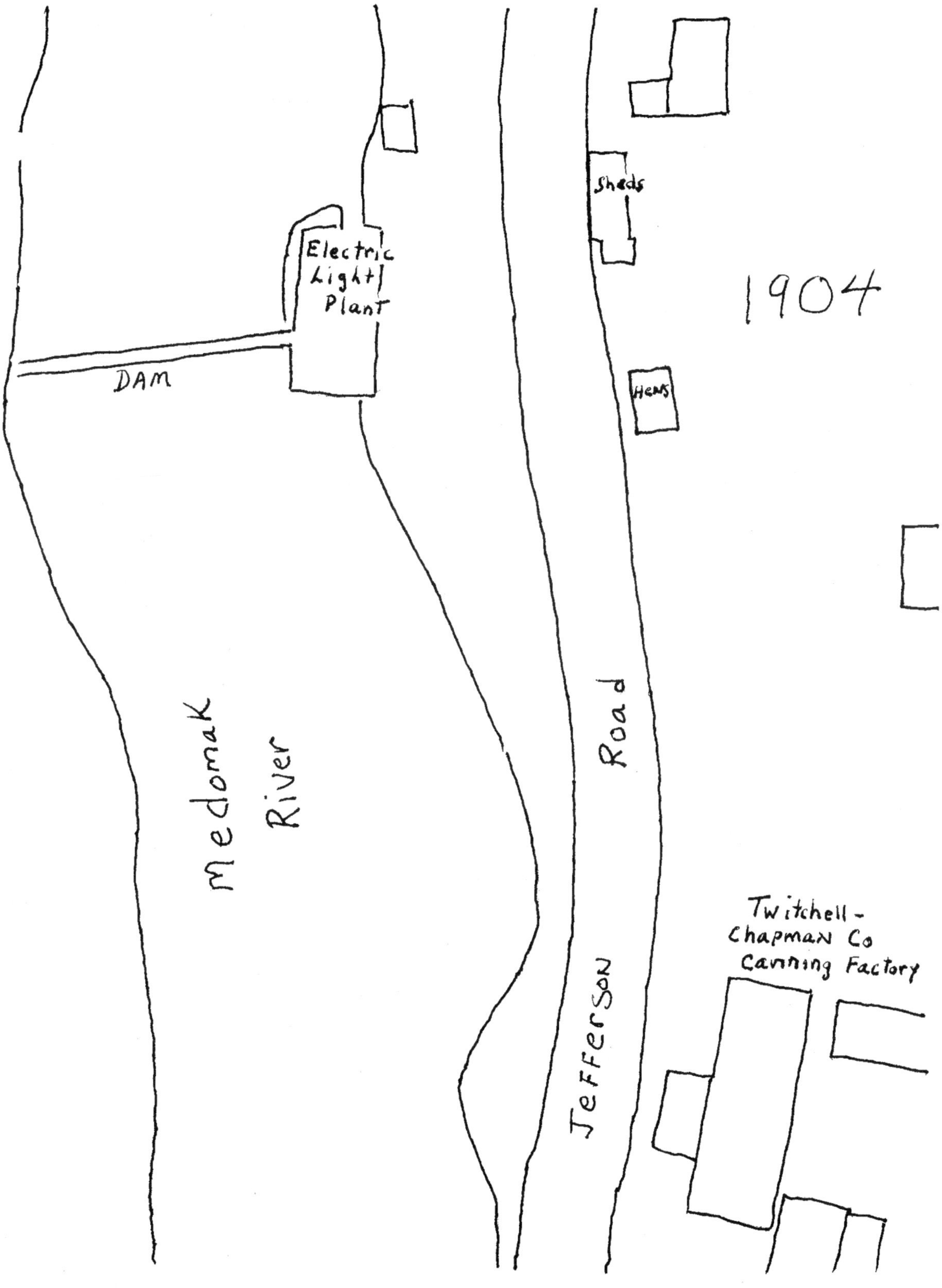

An electricity producing plant arrives, replacing the Medomak Flour Mill.

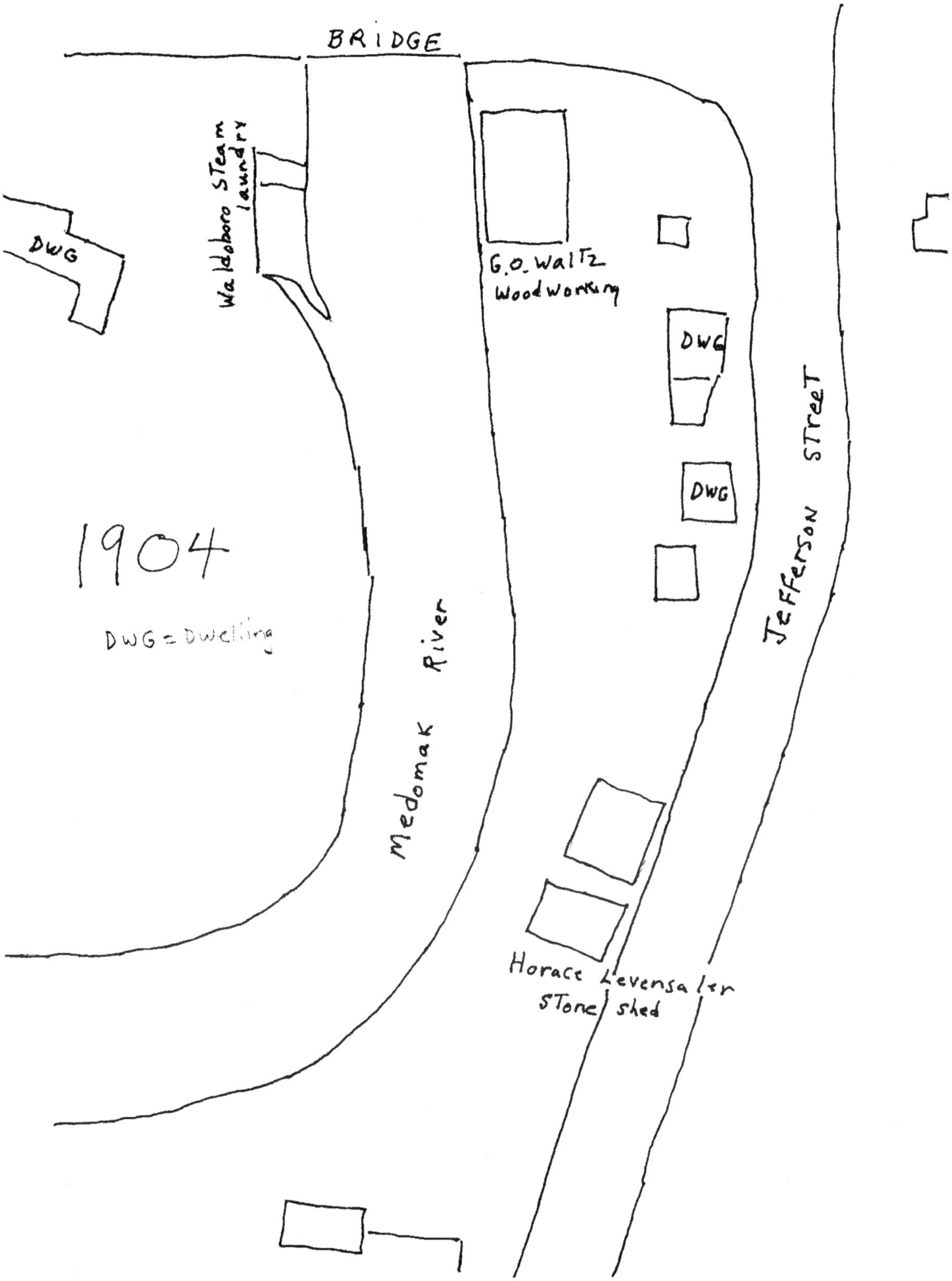
BRIDGE
Waldoboro STeam laundry
DWG
G.O. WalTz Woodworking
DWG
DWG
1904
DWG = Dwelling
Medomak River
Jefferson STreeT
Horace Levensaler STone shed

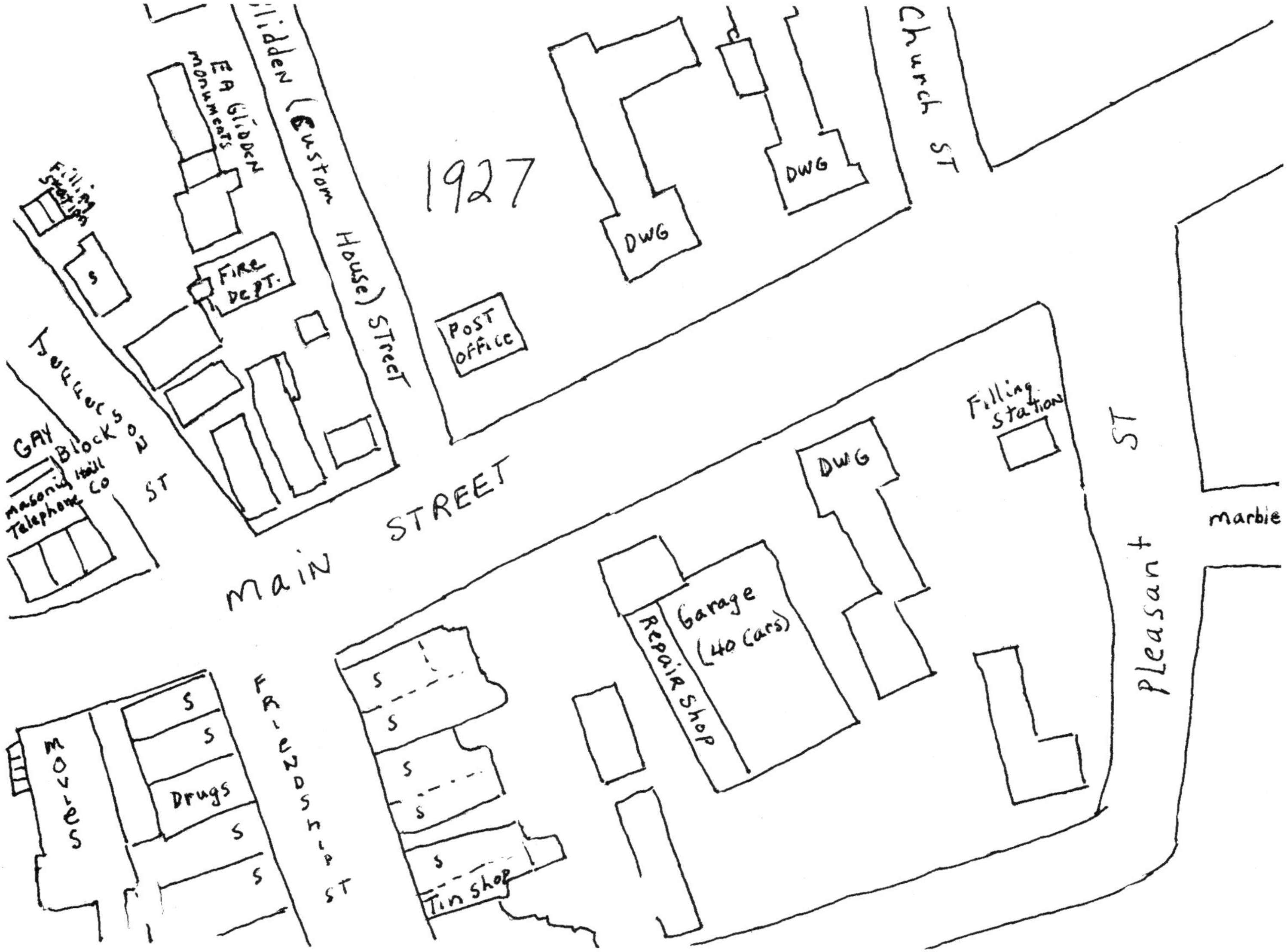

Filling stations, a 40 car garage, a movie theater, and a telephone company!

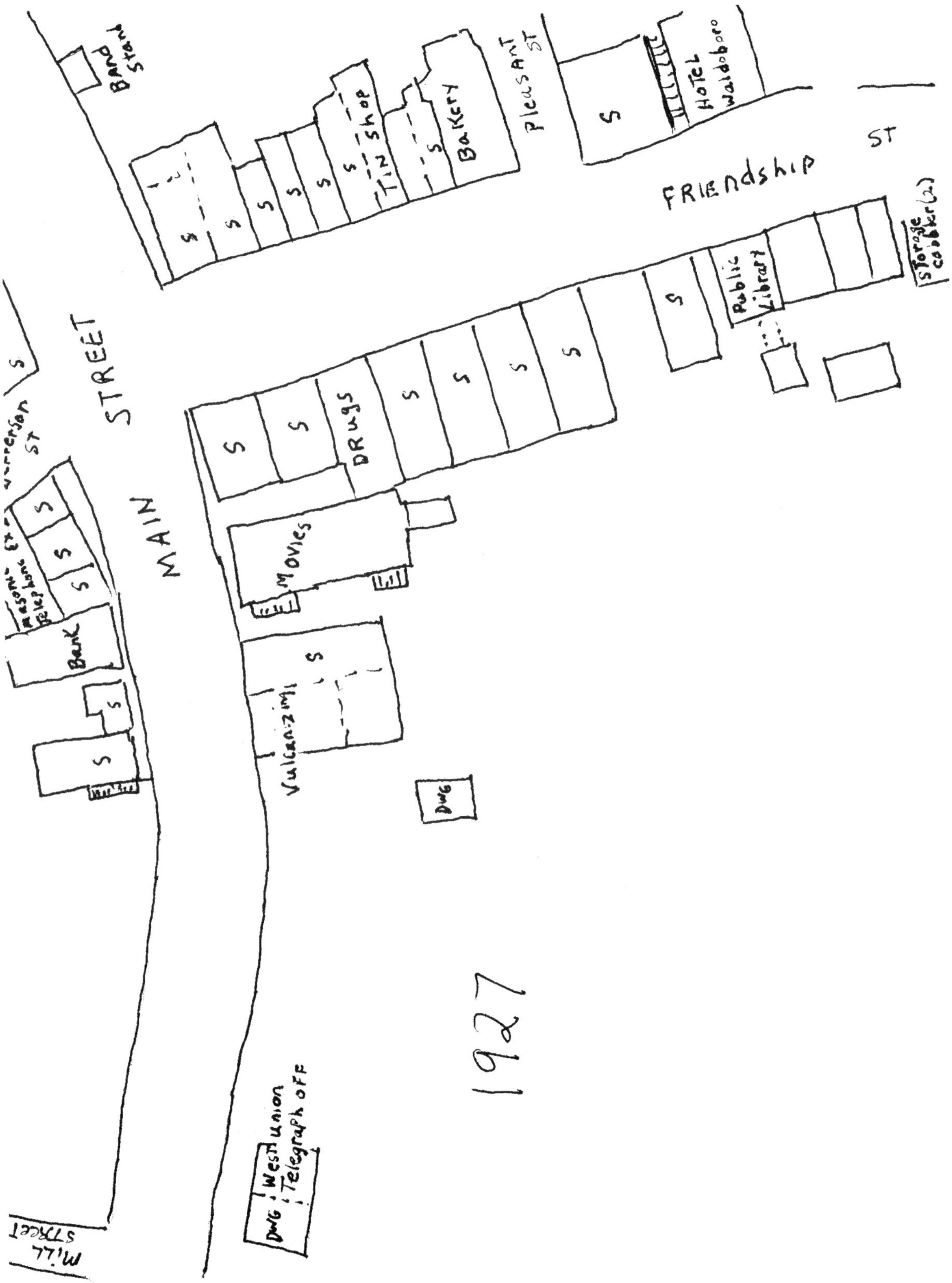
Band Stand
Pleasant St
Hotel Waldoboro
Tin Shop
Bakery
Friendship St
Public Library
Storage
Main Street
Drugs
Movies
Bank
Vulcanizing
DWG
West Union Telegraph Off
Mill Street
1927

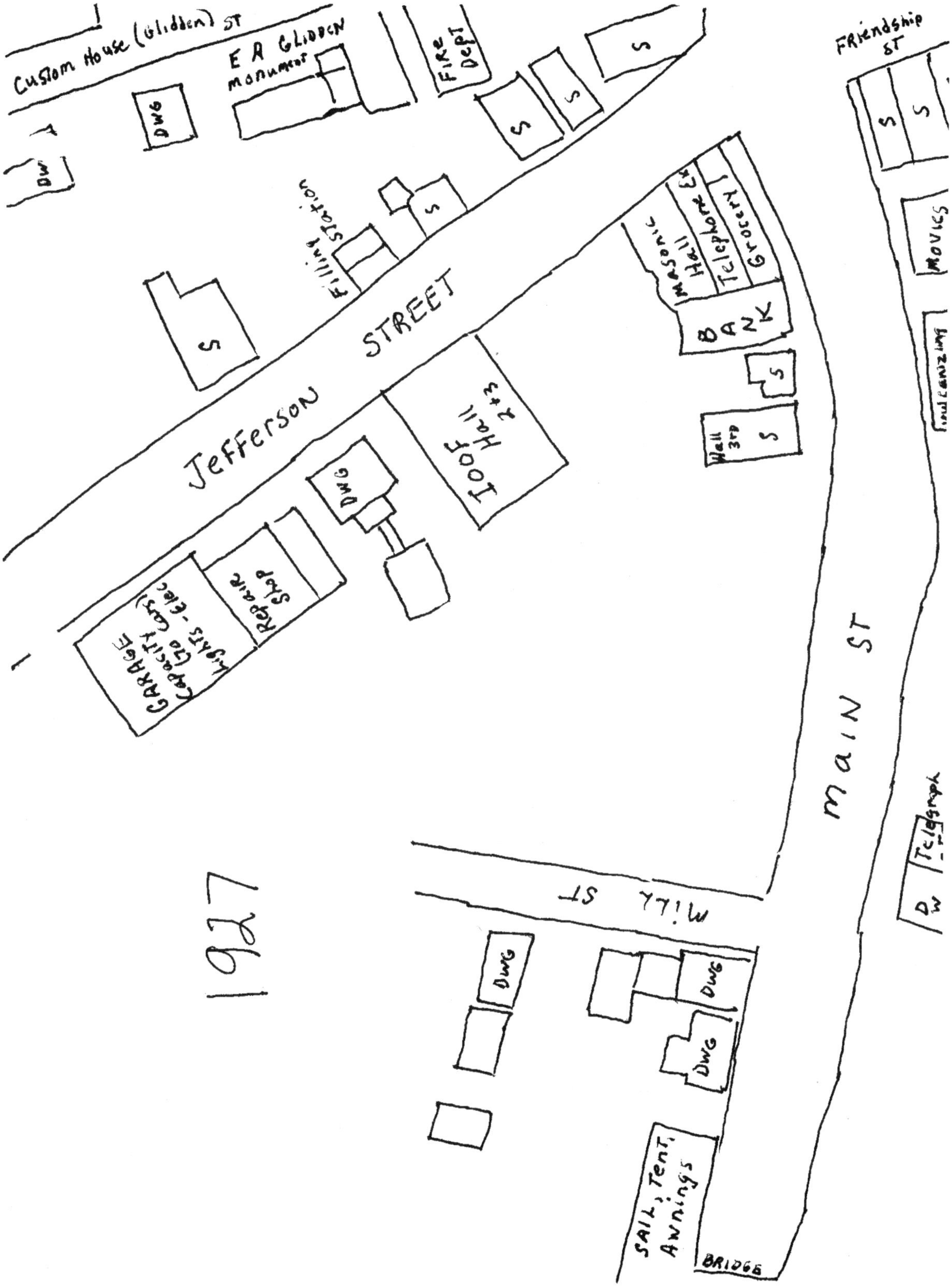
CUSTOM HOUSE (Glidden) ST
E R GLIDDEN monument
FIRE DEPT
DWG
DW
S
FRIENDSHIP ST
MOVIES
Filling Station
Jefferson STREET
Masonic Hall
Telephone Ex
Grocery
BANK
IOOF Hall 2+3
Hall 3rd
Repair Shop
GARAGE
MAIN ST
Telegraph
D W
MILL ST
SAIL, TENT, Awnings
BRIDGE
1927

"Kids, Kids, Kids"

The Lincoln County News, published in Waldoboro, occasionally included anecdotes (and advice) concerning youngsters – what they were doing – and not doing – and what they OUGHT to be doing. What follows are some examples. (We leave comparisons – 1870's – 2000's to the reader.)

January 9, 1875 – "It's GOOD SLEIGHING!!!!"

October 1, 1875 – "A special train will leave for Bath at 5:00 this afternoon for the musical concert. Fare including admission to the concert will be $1.30."

September 3, 1875 – "A boy on Saturday fired a charge through a window in Gilbert Kuhn's house. The blind and the sash were broken, but fortunately no one was hurt as the room was not occupied at the time. Small boys should not be entrusted with firearms. It is reported that several boys in our village carry loaded pistols in their pockets, a rather precocious practice to say the least."

March 19, 1875 – "Rules to teach our daughters – 1. Cook a good meal. 2. Darn stockings and sew buttons. 3. Make bread. 4. Make shirts. 5. Know the mysteries of the kitchen, the dining room and the parlor. 6. Live within your income. 7. Wear calico dresses like a queen. 8. Wear warm, thick shoes. 9. Use hard, punctual common sense. 10. God made us in His image. No amount of tight lacing will improve the model. 11. Have nothing to do with intemperate, dissolute young men. 12. If there is time and money enough, learn and teach music, painting, drawing, etc. 13. Learn to stick to 'YES' and 'NO'. 14. Don't use paint and powder. 15. Regard the morals – not the money – of a beau. 16. Always be truthful, honest, and upright. 17. Remember these teachings – they will constitute a great part of both pleasure and sadness in later life."

April 26, 1877 – "Stacy Clark, son of Edwin, pitched a knife through his foot Friday."

June 21, 1877 – "Seven 'wicked' young men, who attempted to sail in the yacht *W.A. Richard* on Sunday, were nearly capsized by the yacht caught in a jibe. Through the exertions of a retired sea captain who happened to be on board, the sloop was safely made fast to the ship *Willie Reed*, anchored nearby. This ought to be a terrible warning to Sabbath breakers."

February 21, 1878 – "The concert by the Waldoboro Cornet Band at North Waldoboro was a great success. They go next Tuesday to Bremen where they will perform the same concert."

April 21, 1878 – "College students are home on vacation."

December 17, 1875 – "Dear Mr. Editor – 'There has been much complaint of late that the rowdyism exhibited in the lobby of the Post Office in this village upon the arrival of the 6 o'clock mail is an evil which ought to be stopped. Boys and girls fill the place and keep up a continuous uproar until the mail is distributed. Is this a good place for girls to improve their morals? Parents, if you send your daughters to the Office, wait a half an hour and avoid the bad associations to be met during mail distribution. It would seem as though the Post Master, or somebody, ought to be required to enforce better order during the short time that the crowd occupies the lobby.'"

February 28, 1878 – "On Saturday there was a fine sociable for the young people at the Congregational Church."

September 23, 1878 – "No less than 20 Damariscotta boys were in our village Sunday afternoon and evening. Rumor says that some of our young ladies are the attraction, and we would agree from all appearances on the streets Sunday evening."

July 4, 1878 – "A baseball contest between the 'Modocs' of this place and the 'Lightfoots' from Nobleboro was played on Sproul's Level. The result was a 14-11 victory for the latter club. There will be a rematch, we are sure."

A Pickup Game – Author's Collection

Waldoboro Baseball Players on Sproul's Level – Note the two harpists
Courtesy Waldoborough Historical Society

Front row on the ground: J.P. Vannah, George Welt
Second row on the ground: Harry Clayes, Da. Johnson, Mel Head, Clint Stahl, Warren Sanborn, Jack Frost, Ned Bliss, Bill Crowell, Myles Standish
Standing: Elijah Levensaler, Press Whitney, Wash Levensaler, Duke Winslow, John Overlock, (two musicians unknown), John Stahl, Phon Larabee, Isaac Eugley, Os Head
Standing in the rear: Myles Castner

The Lincoln County News – October 19, 1899 – "Waldoboro's baseball team won the first game of the season over Round Pond 17-13. After the game the boys marched up to Union Church where they had supper. After supper they gave members of the circle three round cheers. Then they marched from the church to the Hotel de Harbor View where music was enjoyed until the dance began. Everyone hurried to the Hall and had an old fashioned Social which lasted until midnight. Early in the morning they started back for Waldoboro. All voted it a good time."

Playing Baseball on Sproul's Level – 1927–1929

J.S. "Stan" Hanna

Note – In doing research on Waldoboro in the old editions of *The Lincoln County News*, I kept running across accounts of Town baseball teams playing ball on "Sproul's Level". Not having any idea where Sproul's Level was, I inquired of several people who had lived here for many years. My usual sources "struck out". Finally I was informed that "Stan Hanna would probably know"; I called, him, introduced myself and was told, in a most cheerful way, that not only did he know but he would take me there! After we visited the old field, I asked whether or not he would be willing to write about it. Happily, here is his account.

"The information that I have obtained about this field, I received from two generations of men who played here – before my time. The late John Miller and Earl (Dick) Benner were teammates in High School from 1911–1914. From them, I learned that around 1890 (note – maybe before that) a number of local boys selected this cow pasture to play baseball in. The reason they did was that most people raised livestock, as well as crops, and they did not want to sacrifice any cultivated land for a ball field.

The right way to reach Sproul's Level is to go up Marble Avenue to the dead end and then to a logging road (at that time). At first you start up a very steep hill for about 50 yards, then go south for about 100 yards, then go east on another steep grade to the top. I would say that this is one of the highest elevations in Waldoboro.

In 1927, I was in the 8th grade and playing baseball on the high school team. For our first game in May we all trudged up that hill with two baseballs and two bats. As we approached the top near the third base line, we could see that the field was anything but level – in fact, it had two levels. Because of these two levels, the third baseman could only see the head and upper body of the first baseman. The first baseman had a good position as the base was right next to a big oak tree that he could lean against. The catcher's box (my position) was three feet in front of a stone wall, but you could see over the entire field.

On balls hit hard outfielders and infielders had it rough, having to run around rocks, juniper bushes, and small trees. After a few weeks, one of the fellows brought along a brush cutter to go with his glove!

In early spring people turned their cows out to pasture, so before we got to play ball, the cows had run all over the place. Needless to say, any base runner attempting to steal or slide into a base had to make certain that the base was the right thing. I can say now that there were a few mistakes made; I can recall base runners crossing home plate near me who didn't look or smell too good!

Living in East Waldoboro in those times, my standard equipment going to school was a baseball glove, a shot gun, and a lunch box – in that order of importance! And Sproul's Level was very convenient for me. After a game in the Fall, I would drop down the other side of the hill, pick up my shot gun and hunt birds on the way home!

In closing, I would be disrespectful if I did not mention our High School principal and coach, Mr. Earl Spear, one of the finest men and teachers that Waldoboro ever had."

Sproul's Level Forever Remembered,
Stan Hanna

Through a Teacher's Eyes 1847 (Part 1)

Albion P. Oakes is best remembered as an accomplished Waldoboro lawyer who practiced with the Honorable S.S. Marble throughout the 1850's. He is known also to have married Ella Clark, one of master shipbuilder Joseph Clark's daughters, and with her enjoying the busy social life of that famous family. But years before all this success, while finishing up his law degree at Waterville College (later Colby), he made the acquaintance of one Alden Jackson. This Waldoboro gentleman had taught school there with great success before going into the law. Because at the time Albion was very short of money, Alden suggested he drop out of college and teach for a while; he recommended his home town as a place to start. Albion hade a college degree, having graduated from Waterville at 18 years of age; he would have to gain teaching experience. Vacation time could be spent continuing with his studies for the law degree. (Author's note – Alden would eventually marry ANOTHER of Joseph Clark's daughters, Caroline, further cementing the two men's friendship.)

In May of 1847 Albion Oakes, 21 years of age, arrived in town. He had come from Augusta by stage as he clambered out at the Medomak House Hotel on main Street "thumped about" and tired, he thought about the three formidable tasks ahead of him. First and foremost was to obtain the Village school job; he would be competing with other candidates but was confident about his chances. Secondly, he had to find a place to live – teachers usually boarded at the houses of townsfolk. Thirdly, when school opened for the summer term, he would have to face the students on their home turf.

Albion wrote many letters home to his brother, Abner. Thanks to Mr. and Mrs. William Travers who obtained the letters some time ago we are able to learn much about this young man, his understandable feelings of homesickness, anxiety, and apprehension (at first – quite a lot of self confidence later) about the booming town of Waldoboro in 1847 whose children he was wanting to educate.

1. Obtaining the job – "After I had taken my dinner, I called on Mr. Balch and Reverend Walker, two members of the Committee. There were also four trembling young ladies (candidates as well) in the examination room. I was soon under the screws but had an easy time of it, for the Committee seemed to prefer me as a teacher, taking it for granted that I knew as much or more than either of them. I went through the examination with Honor to myself, and shortly thereafter received the certificate vouching for my acquired abilities to teach the good scholars of Waldoboro."

2. Finding a place to live – "I inquired as to where they were going to send me to board to which they replied that I should go to Mr. Willett's FIRST. Great God!! The meaning of that one word I learned – it was the custom for all teachers to "board around" through the whole village – one or two days at each house. THIS is to repay me for my damn folly in coming here at all! I don't WANT to "board around!" I made up my mind at first to label my trunks for Waterville, put them back on the stage, and return as far as Augusta on foot myself, as I had not enough money left to pay my way back by stages and boats. But finally I made up my mind that I would stay – for a much shorter length of time; they had wanted four and one half months – I would give them two and one half weeks! If I could board at ONE place

only, I would gladly extend my stay. But this keeping school for $20.00 a month and boarding around is not what you or I would like at the midsummer of life!" (Note – After proving himself to be an excellent teacher and a firm but just disciplinarian after two weeks in school, Mr. Oakes was approached by the Committee and offered a permanent room at the Medomak House Hotel for $3.00 a week paid for by the town. "In less than two hours I was pleasantly situated in one of the finest rooms in the tavern, living high. I am emphatically a gentleman upon the town!")

Next – Albion Oakes meets the Waldoboro Village scholars.

The Augusta stage arrives in Waldoboro at the Medomak House
1973 Waldoboro Bicentennial Book

Through a Teacher's Eyes – 1847 (Part 2)

In May of 1847 21 year old Mr. Albion Oakes arrived from Waterville to apply for the vacant teacher's position in the Village School. This wooden structure was located on Friendship Road and was one of 31 school districts to be found in Waldoboro through the 19th and early 20th centuries. Oakes, in the midst of studying at Waterville College (later Colby) to become a lawyer, was taking some time off in which to earn some much needed money. (Later he would become a lawyer in Waldoboro, a partner of the Honorable S.S. Marble; he would also fall in love with and marry Miss Ella Clark, daughter of Joseph Clark, millionaire shipbuilder of the town.) This account, however, tells of his struggles to survive the students of his district as a completely inexperienced schoolmaster. In letters home to his brother Abner he describes these struggles.

May 25, 1847 – "I went into school Monday morning and commenced operations. Eighty or 90 scholars between the ages of 10 and 16 greeted me and proved to have been under the worst management I ever had met within my short life. They are the most saucy and independent set you ever saw, having been bred here in the street among sailors and ships' carpenters. And to make things doubly bad, the man who had kept school here the past year was a "do little" of a thing, the results being that many of the students are accustomed to playing the fool in the street and the master in the schoolhouse. But if you could have been where you could have seen me knock some of them about, you would have thought of our old teacher Mr. Williams. The scholars thought of me as the Devil himself, but if I can't put them in a way to remember me after I'm gone, then what good am I?

The Committee is coming to watch me next Thursday, and I must do well. This is an almighty hard school I am in, and I am homesick. What is the news? Write as soon as you receive this and let me know if you have received that money from Father. I probably will not have the time to continue my (law) studies this summer. If you fail to write as soon as you get this, I'll never forgive it in you!"

But after two weeks of teaching, Albion's success had mollified his homesick thoughts, and they were replaced by feelings of self confidence. His abilities in being able to "govern" his school had made him paramount in the opinion of the villages. He was assigned by The Committee to a permanent room at the local Medomak House Hotel; he began to make the acquaintances of some young ladies and gentlemen. He was now MUCH more sure of himself!

June 5, 1847 – "I received from you a letter full of consolation to your (supposedly) homesick brother. Your advice and counsel were willingly received, but the most acceptable thing of all was that five dollar bill which proved a cure to all mental diseases. I have been teaching now two weeks with first rate success. Those people who, after the commencement of the term, would not send their brats to me for fear that I should kill them, have come to the conclusion that I know how to keep school better than they do, and that I should not kill anyone in cool blood; they therefore have ventured to send their trembling babies to the place where I reign King. There were the second week some 10 or 12 who dared not send for the fear that I might kill their children if by chance they made the least noise, and who changed their minds after being told by me that I would not hurt anyone, as long as they behaved themselves. They now trust me.

After I had kept school about a week, the committee gave me a call which resulted in the report that I was the best schoolmaster ever in the town of Waldoboro. I have got my school so that they obey en masse at the nod, are prompt at the hour, are still in school and noiseless out – a fact that confounded all the old women of the place at first!"

Two weeks more and he speaks of academic successes as well:

June 21, 1847 – "Now all is satisfaction and peace. My scholars are studying hard and learning fast. I had two little girls close (complete) their Algebra this afternoon, having been only 23 days going through it, which is believed by the district to be unparalleled. I am now considered the best teacher ever in the place – Jackson (Albion's mentor in Waterville who had taught in Waldoboro and was responsible for finding him the job) excepted – he (Jackson) is thought of almost as a god! I have been very steady and straight forward in my business, paying no attention to what was going on beyond the schoolhouse, but THERE being the MASTER in every sense of the word. My being able to make sure that everything was proper and that everything was done better than necessary by the scholars both in and out of school without any trouble has caused considerable wonder among the people of the place."

The Agent (responsible for the payment of the teacher in his district) visited Albion and offered him "more than I could get from any school in the state" (presumably more than the $20.00 a month that he was receiving). But Albion did not promise he would remain beyond the seven week summer term – he wanted to be closer to home and his college law studies. He took pride, however, in the fact that, for the moment, he had conquered Waldoboro!

Brick Schoolhouse with students – Courtesy Waldoborough Historical Society

The Lincoln County News – June 27, 1902 – Graduation exercises for the Waldoboro High School Class of '02 were held at the Congregational Church. The motto for the class is: "Climb Though the Rocks Be Rugged."

School Days, School Days

The accounts of Albion Oakes' encounters as a fledging school master interested us in the history of education in Waldoboro up through the early 20th century. For our information we turned to Stahl's History and also to an essay written by Barbara Pitcher of Waldoboro High School's Class of 1934.

In the very early days education was not a high priority in the town. From 1773–1785 all attempts to raise money for schools failed totally; finally the towns folk from 1785 to 1795 voted to spend $200.00 to equip ALL of the scholastic needs for all buildings and equipment. Despite Waldoboro's growing prosperity because of shipping and shipbuilding, most people did not have extra money to spend, and those who had it felt reluctant to spend it in this area. As a result, sadly for the children, the schools were on the whole small, unsanitary, poorly ventilated and heated.

As Albion Oakes discovered, Waldoboro in the 19th and early 20th centuries was divided into 31 school districts each of which had a building located within (sometimes long) walking distances of the scholar's homes. An agent would negotiate the teacher's salary for either the summer or the winter term – each from 12 to 15 weeks. Wages averaged about $20.00 a month plus about $3.00 for board. If the teacher were "good" – could keep the students in line as well as make them learn, he or she might be hired for another term. This rather haphazard system, plus the habit of hiring totally inexperienced masters (like Mr. Oakes) encouraged the growth of alternative schools and private tutors for those children whose families could afford it.

In Waldoboro Village (District No. 6) the school that Albion Oakes learned to "govern" was replaced in 1857 by a brick structure we know today as the Friendship Street School. Voters of the district, buoyed by shipbuilding profits, voted to spent $8,000.00 on this project; it was for the Villagers only – others wanting to attend had to pay tuition. By 1886, however, the financial burden was too great, and the entire town voted to take over the building, open it to all and call it Waldoboro Free High School.

This high school's history extended into the 20th century with several other important official moments to remember:
1879 – A sum of money was raised to buy a large bell to ring the students to class. (One legend told to me by the late Mr. Frank Boggs describes the trick of tipping the bell up, filling it with water, and waiting for someone to attempt the ringing of it unknowing of the sudden bath they were about to receive. To delight the tricksters even more, the water occasionally froze solid in cold weather.)
1894 – A permanent assistant teacher was appointed.
1914 – The "School Improvement League" organized volunteer work on the building and grounds. $375.00 was raised; of that sum $215.00 was spent on the new Kennedy Baseball Field adjacent to the school.
1917 – After a great controversy, the citizens voted a large sum of money for a heating system. Some thought it wasteful and unnecessary, but in later years the vast majority were happy about the decision.
1923 – A third teacher was hired.

1927 – The Commercial Course of study was added to the curriculum and turned out to be very popular. (Most students did not go on to higher educational institutions even after graduating; instead they tried to find work in town – more and more difficult as the economy worsened.)
1928 – An Athletic Association was formed and as a result, progress was made in the quality of baseball and basketball teams.
1929 – A fourth teacher was employed. And a fifth teacher to help with the addition of the 7th and 8th grade students added to the school at this time – there were now 150 scholars in total.
1934 – Students were now able to choose the College, the Commercial, or the General Courses of study. If they wanted to attend college, the school for the first time was certified by the New England College Entrance Board, and they would no longer have to take an extra examination.
However, the crowded conditions (three classrooms and a closet for typing) made it necessary to build a new high school.
1936 – Waldoboro High School (now A.D. Gray Junior High) was constructed.

A List of the Town's School Districts
District number
1. Dutch Neck 2. West Waldoboro 3. Kelly District (midway between Waldoboro and Winslow's Mills, west side) 4. Hoch District (east side between Winslow's Mills and Orff's Corner) 5. Orff's Corner (near Mr. Harold Achorn's) 6. The Village 7. Winslow's Mills (east side) 8. Winslow's Mills (west side near factory) 9. Hahn District (east side, midway between village and Feyler's Corner) 10. Feyler's Corner 11. North Waldoboro 12. "Old Proctorville" (on crossroad connecting road above Winslow's Mills east side to North Waldoboro Road at Davis Corner) 13. Chapel Corner 14. Nobleboro on Route One 15. Manktown 16. Goshen 17. Slaigo 18. Farnsworth 19. Back Cove 20. Ledge School 21. Finntown 22. Genthner neighborhood 23. Old Hoch Road leading across to Feyler's Corner 24. Bremen's Broad Cove 25. Gross Neck 26. Orff's Corner (near the late Albion Achorn's) 27. Levensaler or Belscop School 28. Weavertown 29. Kaler's Corner 30. Flander's Corner 31. Bogue's Corner near Teague's

— This list was taken from a report made by the U.S. Geological Survey in 1915 and on data furnished by Miss Agnes Creamer of Waldoboro.

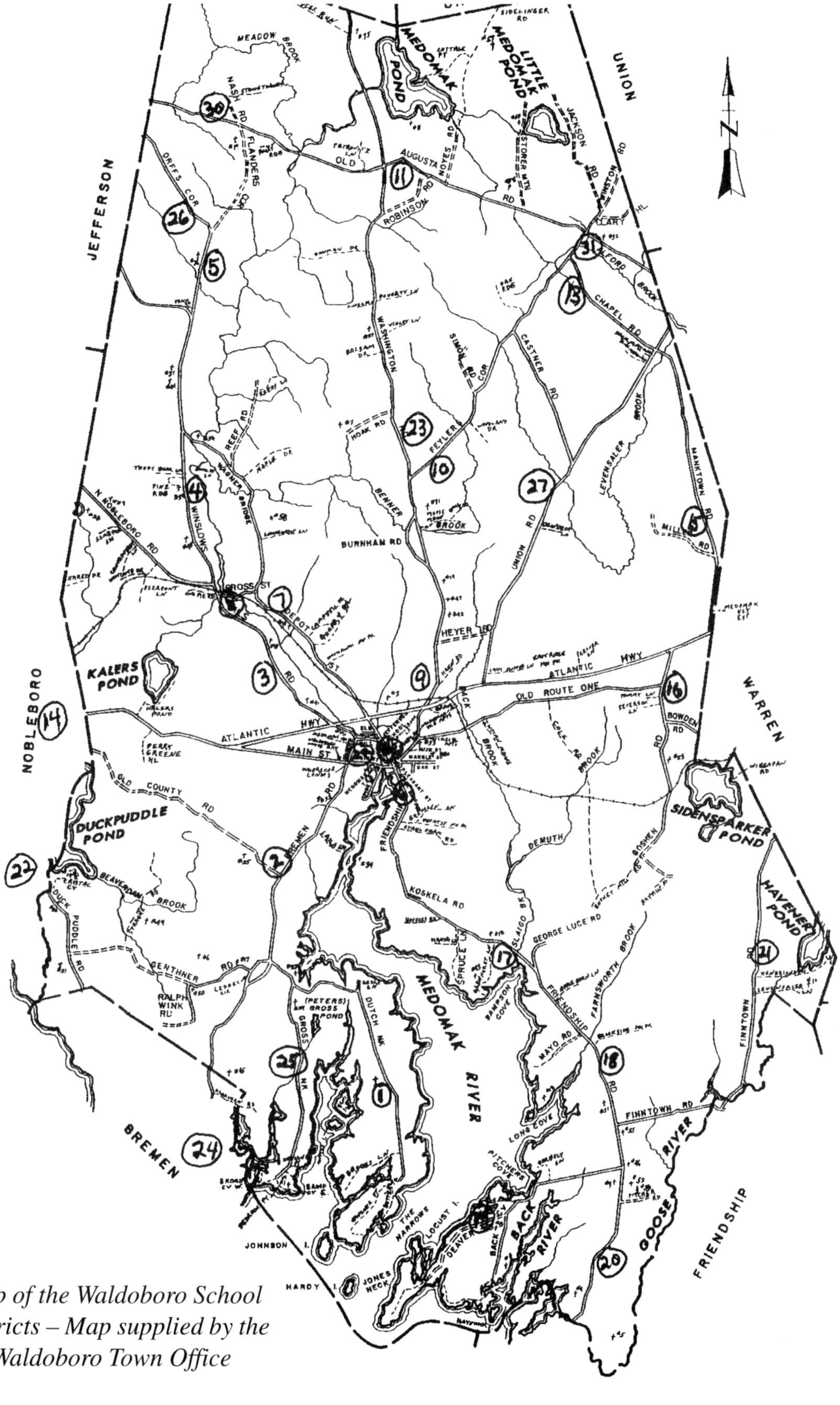

Map of the Waldoboro School Districts – Map supplied by the Waldoboro Town Office

To Go For the Gold – Or Not (Part 1)

(This story is derived from the letters of Albion Oakes (1826–1859), a Waldoboro schoolteacher and later lawyer. Many thanks once again to Mr. and Mrs. William Travers for the use of the letters.)

By the summer of 1849 Albion Oakes, aged 23, had, after a somewhat rough start proven himself to be an extremely competent and popular schoolmaster. So well thought of was he that he was asked by the School Committee to assist the members in the choosing and tutoring of new candidates. (One candidate was his younger brother Abner.) While torn in his mind over whether to continue his own teaching career in Waldoboro, look for a higher paying job in another town (Cony Academy in Augusta was a possibility), or break off education completely and resume his law school studies, he was convinced by townspeople to stay.

Albion had established lofty goals for himself: "Be a success, make money, rise above the common folk." Unfortunately, teaching school at $30.00 a month was not likely to satisfy goal #2; he forever worried about finances and was forced to ask his brother and his father for money to purchase such every day items as underwear. He purchased school supplies with his own savings. And he was also carrying on – and enjoying immensely – a heavy social schedule (although becoming more and more interested in Ella Clark, daughter of the wealthiest Waldoboro shipbuilder, he was attending parties three or four times a week). The Clark and Haskell mansions in the middle of the village were busy in the evenings; Albion remarked in one letter to his brother that he was "courting six girls at once in the village". What could he do about more money to help him be a success and rise above the common folk? And then an idea began to take root.

In January of 1848 gold was discovered in California. For men all over the United States the fever of becoming rich quickly was hitting hard. Waldoboro was at its peak in shipbuilding, and some of those vessels were destined for the West. Joseph Clark's brig *Caroline Clark* launched in November 1849 was one of these. Albion was bitten by the Gold Bug.

In a letter to brother Abner October 25, 1849 he declared:

"I will commence with the subject of an expedition to California. Your object and mine in studying and practicing law is to make money – to get rich – to demand influence, and to live easy. But if we are lucky beyond the masses, it will require years of hard study, deprivation, hard labor, troubled conscience, and aching heads before we can expect event to possess a competency. If we practice law, our life will be one of continued strife and litigation, IF we obtain our license and pay our expenses over the next five years. BUT should we pack up and go to where money is to be had, we stand a much better chance of becoming rich. Many people are becoming more than wealthy; perhaps you recollect a fellow who lived down in the field on Friendship Road this side of Captain Sampson's – the fellow never done a good day's work in his life? Well now after nine months in California he has $100,000. Now THAT can't be earned practicing law! (Author's note – teaching school as an alternative is not even mentioned!)

"They say that there will be more than 300 vessels from New England next month, and they will carry a mighty host of men. What I propose is this: We get a group ready to go between now and the end of December and go by way of the Isthmus (of Panama) rather than by Cape Horn; this will save thirty or forty days. We'll need money, and the way I propose is let someone else pay for the expedition in exchange for a share of the profits.

"I don't know of anyone else who would stand a better chance of making money there than we should, for we are both young bloods, strong, and rugged. We go with steady determination to work hard, watch the chances, and are ready to take advantage of everything. We can teach school, tend a store, survey lumber or land, or, if need be, preach. In a word, we are Yankees! I am confident of one thing – that if we stay here, we shall never become rich, but if we go, we stand a possible chance. Nothing risked, nothing gained. I hope you conclude to go, but I shall go alone if I must."

(To be continued)

Joseph Clark Mansion, Pleasant Street (also known as Dog Lane); grandchildren in the picture include Joseph Jr., Stacy, and Mabel – all sons and daughters of Joseph's son Edwin.

Morc About Albion Oakes (from his letters 1847–1850)

When we last discussed Albion Oakes, the young Waldoboro teacher was trying to decide in the winter of 1850 about the possibility of going to California to gain riches from the gold fields. He had written to his father who emphatically discouraged the idea; his brother Abner, on the other hand, also a Waldoboro teacher, was very generous with financial support. In a letter to Abner and another brother, Valentine, he told of a local "company", headed by men named Genthner, Harriman, Morse, and two others who planned to leave for the fields in April; they urged him to come when he could, promising him a partnership when he arrived. The price of passage was $425.00 in April but would hopefully be less in the summer. Albion decided to wait until then. "When they arrive in San Francisco, they will write me how things look and what the prospects seem to be, and from there they will go directly to the mines and commence operations, and as soon as they get under way so that hey can write with certainty about the matter, they will write again, giving me full and direct account of things as they find them. If (things) are good, they will write for me to come, and if not, to stay where I am. They also promise me, if they find things right, and money is to be had for the labor, that they will write me, and if I will come, they will take me right straight into the company on equal shares – the same as if I had gone with them. This would seem to be the best course for me – with a smaller risk – and a smaller passage fee. I am leaving it this way for the present.

I am much obliged to you for offering me all your money to help me off, but as it is, I shall not want it this spring and shall probably want your help in the summer if I go, as Father perhaps will not (help). There is money to be made in California, and I am bound to have my share of it. If I do go and make money, you shall not lose anything by being so willing to shell out your last dollar to help me."

Unfortunately (or fortunately – from his father's viewpoint) Albion was never able to venture out to California. Perhaps it was too much of a dream – and too much money. He decide to stay and try hard to make a living, after all. Teaching the young people in Waldoboro had been very successful; he was employed to instruct 56 prospective eductors at a Teachers' Institute in town. He applied unsuccessfully for a job at a girls' high school in Bangor in April of 1850 and at the high school in Augusta; by August of 1850 he landed one in Gardiner. Later that year he taught a term in Winslow. But he continued to "look for something better – which would also pay him more."

He returned to Waldoboro after the fall school term and visited with his friends. "I lived mighty fast while there." Having studied and read law off and on and urged to go into it full time he was offered a job locally with John H. Kennedy and another with E.B. French. Eventually he was to become a partner with another Waldoboro lawyer, the Honorable S.S. Marble.

For Albion, there was another attraction in Waldoboro which was becoming more and more important to him. Ella Clark, daughter of the mighty shipbuilder, Joseph, had recently graduated in June of 1850 from Gorham College with High Honors in Scholarship and Composition. Albion had known Ella through her older sister, Caroline, who was to marry his college friend and teaching mentor, Alden Jackson. He had known Ella socially at the many Pleasant Street parties attended by the young crowd in his first year of teaching.

Albion traveled that June day to Portland "on the cars" to attend Ella's graduation. He was two months short of his 24th birthday.

Sources for this story: Miller's *History of Waldoboro*, The letters of Albion Stahl, Lubbock's *The Down Easters*.

Ella and Albion (last in a series)

(Celeste Clark, the youngest of shipbuilder Joseph Clark's daughters, lived from 1837 to 1929. She was witness to an incredible passage of historical events – in Waldoboro and the world. She watched the terrible fires of 1854, 1892, 1893, and 1900 which transformed the face of the town. She watched young men march off to the Civil War, the Spanish American War, and World War I. She witnessed the inventions of the telephone, telegraph, the automobile, the radio, and all the usages for electricity. And she saw the demise of wooden shipbuilding, her father's and Waldoboro's lifeblood. Waldoboro, in her years was transformed from a bustling center of business and trade to a sleepy depressed hamlet with a huge decline in population. In this span of time she outlived, by many years, all her sisters and brothers. We will pretend that she is writing this story.)

I, Celeste Clark, was a girl of 10 when the young schoolmaster, Albion Oakes came to town in May of 1847. It wasn't long before I overheard my parents discussing him and his remarkable success as a teacher – making students learn while all the time maintaining strict order. My sisters Caroline, 22, Mary, 27, and Ella, 15, were very socially minded and often entertained him and other young people at parties in our big white house on Pleasant Street. Albion used to talk about what he was going to do to make more money – that seemed to be very important to him. At one time he was ready to go by boat to California with some other men from the Village to search for gold, but he never did. I think that he was becoming more and more interested in Ella. Caroline had been spoken for – by Alden Jackson, another teacher who had been responsible for bringing Albion to Waldoboro. Mary was already married – to Mr. William Allen. Ella often claimed that she would NEVER marry, but one could see that Albion was becoming more and more interested in HER as she grew up – prettier and prettier all the time. In June of 1850 we all traveled to Portland to watch Ella receive her college diploma from Gorham State Teachers' College. She was only 17 at the time. Albion came along too. He was 24.

Albion decided soon after the graduation that teaching was no longer for him; he had been studying the law in his spare time and was offered a post with the Honorable Sebastian Marble here in Waldoboro. He continued to court Ella who had become herself a promising writer of prose and poetry and a fine teacher. After three years they became engaged and were married October 17, 1853. Three years later their first child, Carrie, was born, and two years hence another little girl, Minnie, blessed them with her presence. Every thing looked promising for Ella and Albion – they were so happy; unfortunately the happy times were not destined to continue.

Sickness was rampant in Waldoboro. Digestive diseases such as cholera and dysentery, probably caused, we have discovered, by the drinking water polluted by foulness from wells dug on the side of the hills in town, swept away many people; measles, mumps and smallpox also were common killers – particularly the children. I can remember two, three, and four funerals a day. Albion was very sure of himself and of his health – I recall him commenting to his brother; "While others are groaning with pain, I am perfectly well and healthy – first rate – never better." He did have stomach troubles for a few days once, but soon recovered. So it was a terrible shock to him when his own little Minnie caught the fever and died, in June of 1858, having lived only two months. Albion was never the same after Minnie was gone; much of life's joy seemed to have evaporated. And then, one year later, at the age of only 32 Albion himself became ill and died, leaving Ella, at 26, a widow with a three year old daughter to raise.

Albion and Ella had been living in a house Father had built for them (most of us – my brothers and sisters lived nearby) up the street from the large family home. Ella remained in her house for many years but eventually came to live with me and remained as an elderly lady. She continued to teach and write, being known by her acquaintances as a good listener with "an affinity for all that is good and true – and a patriot of a teacher." For many years she, Mary and I were deeply involved in the theater that Father built for us after Mother had died in 1851 and the Great Fire had destroyed so much of the town in 1854. Ella would usually direct the plays while Caroline and I were making costumes and Mary the music.

Ella and Albion's daughter Carrie grew up, married a Mr. Manville of Carbondale, Pennsylvania, and bore a son, Keith. Tragically, Carrie died in 1886 at the age of only 30. But Ella persevered, unbroken, continuing to abide in Waldoboro, our home. She lived out her life as all around her family members were growing old and passing on: Young Joe in 1863, dear Father in 1875, Atherton, the Civil War hero, in 1882, Caroline in 1888, and Mary in 1897. And now, April, 1904 sees the passing of Ella herself. Edwin and I, the two youngest Clarks, remain.

Written April 8, 1904, a few days after my sister's death.

Mark Biscoe's note: The story is based on facts garnered from *The Lincoln County News* – in particular the obituary of Ella Clark Oakes April 7, 1904, The Letters of Albion Oakes to his brother Abner now in the possession of the author, Miller's *History of Waldoboro*, Stahl's *History of Waldoboro*, and an 1857 village map. Surmises occur only in the descriptions of feelings shown by the characters toward each other.

Home Remedies and Health Care

(Source of information – "My Father's Farm" – written in 1981 by Esther S. Gross)

It is only in recent times that Waldoboro citizens have been able to rely on dependable medical service. Now, simply by telephoning 911, a swift ambulance appears almost instantly, ready with trained personnel to assist or to transport the patient to the hospital. In Esther Gross' youth the medical situation had been improved also over the days of the 19th century. Young doctors like George Coombs and the Sanborn brothers had established themselves in town willing to travel in any way they could to the homes of the ill. If snow drifts slowed them down, they simply strapped on snowshoes or skis and ploughed on. But despite this determination and better training given these men, communication and transportation were so slow that most people, particularly those, like Mrs. Gross in the outlying districts, were still relying upon long used and home grown remedies. In this article we will examine some of these remedies; reader, please compare them to the Good Days (medically speaking, anyhow) of the late 20th and 21st century!

For PUNCTURED WOUNDS – a thick slice of fat salt pork bound on with a bandage.
For ITCHING and ALLERGIES – salt and vinegar
For COLDS – bandage of salt pork around neck; drinks – smartweed and water (terrible taste), ginger tea, hot lemonade, and (best of all) – hot toddy with rum.
For INDIGESTION – spoonful of peppermint or wintergreen in a cup of hot water if severe, ginger and baking soda in hot water.
For CHEST CONGESTION – warm goose or hen grease smeared on chest; camphor or petroleum on chest if severe, a poultice of sliced onions. *Note – a poultice is a soft, moist mixture of bread, meal, or herbs spread on a cloth and applied where needed.
For LAME MUSCLES, or RHEUMATISM – Balm of Gilead buds (from a back yard bush) packed in a bottle, covered with pure alcohol and bandages with heavy tape.
For BROKEN BONES – if attended to, splints and bandages with heavy tape.
For CUTS – sewn up at home with needle and thread – no novocaine, of course.
For COUGHS – lemon juice and honey; rum and molasses lard and sugar; sliced onions cooked slowly with molasses and lard; syrup of squills (expectorant, diuretic and cardiac stimulation).

There were also extremely serious (often fatal) diseases which were apt to kill many children:
For DIPHTHERIA – a poultice of flour, salt, and vinegar mixed was spread on a cloth which was bound around the neck of the patient over night.
For WHOOPING COUGH – raw onions in a muslin bag applied warm to the chest and throat. Each day the patient was given a ipecac (to make him or her vomit).
For CHOLERA – adults – one teaspoon full each hour of a mixture of cayenne (red pepper), spirits of camphor, tincture of rhubarb, ounce of brandy, peppermint, small amount of laudanum (a sedative). Children – one half teaspoon every hour.
For PNEUMONIA – chest and back covered with petroleum jelly; then covered with warmed cotton pads; the patient would then be wrapped tightly and tucked into bed.

For MEASLES – if "black eruptions" appeared on the patient's body, it meant that he or she had a good chance to live. Once, in Mrs. Gross' childhood, these eruptions failed to appear on her and her brother. Their uncle was visiting at the time; he went to the sheep pen and collected a pan full of sheep droppings. These he covered with water, mixed with molasses, and cooked up slowly on the stove. This so called "Nanny Plum Tea" was served up hot to the children. It worked! Within a day the black eruptions had appeared and Mrs. Gross and her brother were recovering.

"Nanny Plum Tea" to "Call 911" in the Waldoboro lifetime of one woman!

(The author wishes to thank Mrs. Martha Boggs, Mr. Richard Wallace, and Mr. Randy Gross for assistance in the writing of this article.)

Social Center of Old Waldoboro – Clark's Hall

A few yards from Waldoboro Village's four corners stood for over 100 years a white frame building known as Clark's Hall. Mr. Joseph Clark, the great shipbuilder, owned the land on the south side of Main Street all the way to the Medomak River. In 1855, after the Great Fire had destroyed much of this part of town, he built three wooden properties on the hill: a theater/hall, a business block, and a large house for his daughter, Celeste. (All three of these buildings are gone.)

For many years Clark's Hall was the center for the town's numerous social events. With the theater upstairs and two stores downstairs, it was busy days and as many as five nights a week. Clark's daughters, extremely interested in dramatics, had been the primary impetus for their father's project and could often be found involved in locally produced plays and operas: Ella directing, Mary furnishing music, and Caroline and Celeste designing costumes.

In 1875 Mr. Clark and his son Edwin arranged for extensive renovations and improvements in the Hall: the upstairs theater was rebuilt to be 17 feet high, 36 feet wide, and 55 feet long exclusive of stage and anterooms - 80 feet long over all. Two entrances front and on the side were enlarged. Gas pipes for heat and lights were installed much to the consternation of *The Lincoln County News* reporter who commented: "Gas in Waldoboro? It's too much!" (Later he altered his opinion: "When the outside temperature was 12 below zero, we have NEVER visited a public hall as comfortably warm as Clark's.") After its completion, the building was judged to be one of the most substantial and handsome halls in the state for its size.

It is nearly impossible today for us to imagine life without movies and television. We tend to think of our ancestors who did without these two media as enduring a drab, uninteresting life. Actually, nothing could be further from the truth. Judge for yourself by reading a list of SOME of the entertainment enjoyed at Clark's Hall by Waldoboro and out of town folks. Incidentally, the entrance fees charged were never over a dollar and usually much less!

Feb. 18, 1877 – The Bowdoin College Orchestra will entertain.
April 5, 1877 – Five nights of shows at Clark's this week.
May 30, 1877 – A lecture by Professor H.B. Brown who in 1849 escaped slavery and traveled from Richmond to Philadelphia in a BOX – three feet one inch by two feet by two feet six inches. (!)
Feb. 21, 1878 – Whitman and Clark's Minstrels from San Francisco (note – minstrel shows – both traveling and locally produced were common).
March 13, 1879 – The Waldoboro Cornet Band.
Sept. 4, 1879 – Concert by Miss Luela Winslow's singing group.
Dec. 4, 1879 – Christmas Cantata – local young people.
March 30, 1899 – Drama – Act V, Scene 1, Macbeth – performance by Miss Alice Reed and Maude Clark Gay.
April 20, 1899 – Moving pictures (note – NOT movies!) showing stirring scenes of a Spanish warship being sunk during the recently concluded war. Accompanied by music on a Phonograph.

Jan. 4, 1900 – New Year's Eve Ball – Oyster Dinner at Hotel Davis (next door) – music by Weatherbee's Orchestra of Warren. 35 couples.
Feb. 1, 1900 – Waldoboro Chorus Club – Grand Concert conducted by Professor Stockbridge of Portland.
Feb. 22, 1900 – Historical Pageant by the Senior Class of Waldoboro High School.
Aug. 8, 1901 – Company of Italian glass blowers demonstration.

With the advent of the motion pictures, there was a need for a movie theater in town. Clark's Hall became the Star Theater; many older people today remember the Star with its bowling alley and restaurant downstairs. When the Waldo Theatre was constructed in 1936, business for the Star was drawn away. Finally, during the 1960's, the old Clark's Hall building was demolished – truly the end of an era.

Clark's Hall Playbill – 1882

Jolly Jollier Jolliest Jollities

Return of the Favorites

Clark's Hall – Waldoboro
Saturday, May 27

"FROM OCEAN TO OCEAN"

Houses Packed to the Very Doors
Nothing Like It On the Stage

ATKINSON'S JOLLITIES
In Their
Great Musical Absurdity

Our Own Orchestra
Our Own Scenery
Uproarious Fun

Original and Charming Music

Unbounded Enthusiasm

THE GREAT BURLESQUE DRILL SCENE

Popular Prices
Admission 35 Cents Reserved Seats 50 Cents
Now on sale at Bliss' Bookstore,
Charles Atkinson, Proprietor and Manager

KEEP THIS IN YOUR MEMORY!!!!!!!

The Lincoln County News – After a Jollities performance – "The audience laughed until it cried!"

Clark's Hall front porch – Clint Stahl – second from right; Walter Boggs with tuba. (Local actors and musicians would often put on shows and join in with the traveling professionals. Stahl and Boggs were two of the regulars.) – Author's Collection.

The Lincoln County News – February 7, 1901 – "Valentine Ball at Clark's Hall! The social event of the season! Both ladies and gentlemen are requested to dress to represent a valentine of some kind making the event in reality a masked ball. Masks of all kinds may be obtained at the News office. Music by Gale's Orchestra of Rockland. Supper will be served. Tickets 50 cents; Spectators 25 cents; Ladies in costume free."

Clark's Hall in demolition – early 1960's – Author's Collection

Epilogue

Two scenes of the George L. Welt shipyard – expanded from the days of A.R. Reed
Author's Collection

One Man's Optimism - 1896

A new year – 1896 – was dawning in Waldoboro. Mr. Samuel F. Miller, the forthright editor of *The Lincoln County News*, prepared his weekly column for January 1 with an eye for a yearly summary. Never afraid to lavish praise on as well as criticize the inhabitants or events of his beloved town, he looked back with mixed feelings.

Wooden shipbuilding, the backbone industry of the town for 150 years, was dead. Grass grew in most of the old yards. People were moving away. Fires had destroyed much of the area south and west of Main Street, and there were no private fortunes to help rebuild the burned property as there had been after the great fire of 1854. Realizing all this, Mr. Miller decided to write the following title for his editorial;
"Waldoboro – Past, Present, Future; Are We Progressing or Going Backward?"

Our town, began the editorial, has suffered some "downs" lately. At one time we enjoyed the distinction of building some of the finest ships ever floated. With the substitute of steel and iron for wood we have been forced to the wall. But give great credit to our people who have accommodated so readily to change. It's best to count our blessings; what DO we have?

1. One of the finest shoe shops in the country – employing 300 including 63 locals. (Author's note – This shop opened in 1885, eventually became the Paragon Button Factory, and closed in the 1980's.)
2. M.M. Richards Pants Factory – 19 men and women – 25 "at home" stitchers – producing 10,000 pants, 1500 suits, and 300 coats yearly. (Author's note – This Jefferson Street establishment opened in 1893 and was torn down in the 1920's.)
3. Soule's Steam Laundry – 4 or 5 employees – Mill Street bridge (Author's note – Mr. Soule's father had run a carding mill in the same building for many years.)
4. The Medomak Flour Mill – an 80 X 40 two story building – 1/3 for flour, 2/3 for other businesses – carriages, sashes, doors, etc. Upper Falls, Mill Street. (Author's note – Achorn's Mill opened in 1877, eventually was sold to the Waldoboro Water and Electric Plant.)
5. A productive granite quarry – 50 employees – Depot Street (Author's note – The quarry opened in 1887, supplied many important government buildings with granite and closed in 1910.)
6. Candy (confectionary) factory – 3 or 4 employees – Main Street
7. Bottling establishment
8. Sailmaking Loft – 12 or 13 workers – by Main Street bridge
9. Storer's Lumber Yard
10. Corn and Peas packing plant – corner Mill Street – 25 employees (Author's note – The Twitchell – Champlin plant eventually moved to Greene, Maine in the 1920's having been in Waldoboro since 1888.
11. The Medomak River – a power source for small mills and a transportation route for coal and lumber.
12. An American Express office – Main Street

13. Fine churches, good schools, WCTU temperance society, Dixie Dramatic Club, National Bank, Farmers' Fire Insurance Co., numerous other clubs and organizations.
14. A beautiful bandstand – (Author's note - This was across the street from the current Library.)
15. The Maine Central Railroad – brings freight and passengers several times a day.
16. *The Lincoln County News* – bright, clean, up to date; once a week printing.

What We Do Not Have and Need (1896)

1. Better fire fighting facilities. (Author's note – Antique fire engines had been in use for a long time; a uniformed department had been formed in 1875 but closed down due to lack of interest.)
2. Regular police and police court. (Author's note – The former arrived for the first time in the 1920's.)
3. Development of the Village Center.
4. Better water works. (Author's note – Hydrants first were used in 1908.)
5. More factory – like businesses to replace Sproul's Mill – lost in the fire of 1892.
6. A Savings Bank
7. Street lights
8. Better schools (Author's note – There were over 30 neighborhood schools in Waldoboro at this time. The "new" Waldoboro High School – now A.D. Gray – opened in 1936.)
9. Snow ploughing for sidewalks of the Village. (Author's note – store owners and home owners were responsible for shoveling the snow in front of their places.)
10. Better streets year around. (Author's note – Paved roads gradually began to appear in the 1930's.)

Information for this article was procured from *The Lincoln County News*, Stahl's *History of Old Broad Bay and Waldoboro,* and *Waldoboro*, a collection of essays, published by the Waldoborough Historical Society.

July 27, 1893 – *The Lincoln County News* – "An illustration of the decay of shipbuilding is seen on the West side where the owner of a number of shipyards has been ploughing them up for the purpose of converting them into grass fields. It would be difficult to give a list of the large number of vessels built in those yards in years gone by before the memory of the present generation. The last vessel launched there was by Levitt Storer."

Despite the loss of shipbuilding Waldoboro was still a leading town in the county. With a population of 3145 (down from 4569 in 1860) it was an energetic place with throngs filling the streets – especially on the weekends. There were in 1900 four physicians, two dentists, two drug stores, three bicycle shops, three shoe stores, four clothing stores for men and two for women, a carpet shop, two hardware stores, two candy shops, a musical instrument shop, three barbers, one furniture establishment, two beauty parlors, two pool rooms, and several places to purchase groceries, fish, meat or other provisions, to name not all town enterprises.

After a few years with a flat economy in the area of shipbuilding jobs, Mr. William F. Palmer of Boston suddenly appeared to provide work. Beginning in 1900 and continuing for five years he gave hope and optimism that Waldoboro's "Great Days" were reviving. 100 – 125 workers at the old A.R. Reed yard (now Town Landing) under the leadership of builder George L. Welt constructed six enormous white schooners with promise to build several more. As ever Waldoboro folks looked to the sea, wooden sailing ships and shipbuilders like William F. Palmer.

William F. Palmer (1859 - 1909)
Courtesy: William F. Pamer Collection,
Maine Maritime Museum

George L. Welt (1855 - 1917)
Lincoln County News *photograph*

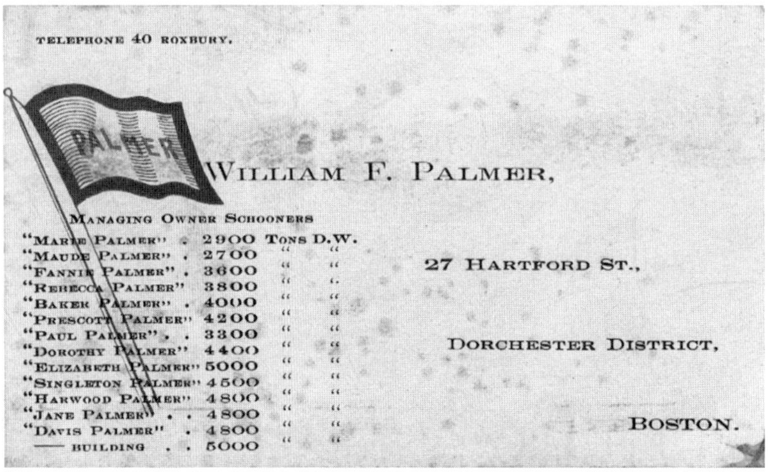

Palmer business card – Author's collection

The Lincoln County News – *September 28, 1899 – "The News is glad to announced that George Welt who is in Boston, has made a contract to build a 5 master here next season for Boston parties – 3 full decks – larger than the Gov. Ames – larger tonnage than any vessel ever built in Waldoboro. Good news to mechanics and citizens of the town."*

Mr. Palmer and His Fleet
Part One

William F. Palmer was the man responsible for the final flurry of wooden sailing ship construction in Waldoboro. Most people have heard of the six 5 masted schooners – *Fannie Palmer, Baker Palmer, Paul Palmer, Dorothy Palmer, Singleton Palmer and Harwood Palmer* – that he had built at the site of the Town Landing between 1900 and 1904. In a later account the reader will hear more specifics about their stories. First, however, it is interesting to learn about the man himself and how he became a ship owner.

For this article this author procured information from the files of the Maine Maritime Museum and from the book *A Shipyard in Maine* by Ralph Linwood Snow and Captain Douglas K. Lee.

William F. Palmer was born in 1859 and grew up, far from the saltwater, in Webster, Massachusetts. His father was killed at the Civil War Battle of The Wilderness in 1864. After graduating from Williams College in 1880 and earning a Master's Degree from the same institution in 1883, he became an educator and for the next 16 years served as teacher and administrator in various high schools – public and private – in Massachusetts and Rhode Island. While at Bristol Academy in Taunton, Mass. from 1888 to 1898 he apparently studied some naval architecture, developed a strong interest in the subject and began to design yachts, two of which he later had built for his own use. His first wife having died, he married again – Marie Elizabeth Convers, later his business associate, who may have influenced him in his decision to change careers.

During the time Palmer was at Bristol Academy, Taunton was becoming the main distribution point in New England for soft coal (bituminous), the primary fuel in those days for homes, industries, and railroads. Bringing the coal to New England was often carried out by the sea. Later on Boston replaced Taunton as the distribution hub.

In 1898 Palmer accepted an offer to become principal of Malden, Mass. High School. After being there only one year, he resigned his job, bought an unfinished Bath vessel, and within months launched the first of 15 large schooners (built at four different yards in Bath, Waldoboro, Boston, and Rockland.) There were two 4 masters and thirteen 5 masters. They were all painted white and they all carried the same houseflag pattern – a rectangular white pennant outlined in blue and with large red letters. All had the name ______ *Palmer* - either for family members like Paul (his son) and Marie (his wife), or for large benefactors such as Harwood or Singleton. Their primary business was to be coal carriers.

Palmer was self-assured and very confident of his own abilities and possessed strongly articulated views on a wide range of subjects. He was perfectly happy to serve as managing owner, designer, recruiter of capital, construction supervisor, and handler of day to day operations. He was a gadfly, writing innumerable letters to his builders, his captains, and his shareholders and he micromanaged every facet of the business. (Author's note – We have Palmer's letters to read; unfortunately we do not have the answers from his probably beleaguered associates!) But business was VERY good because of, or in spite of Palmer. From 1900 to 1907 the fleet showed a 20% profit increase each year. In 1905 the schooners made 40 voyages to Boston and Portland, Maine carrying coal. 1906 showed the *Davis Palmer* alone making 13 trips. When coal cargoes could not be obtained, vessels carried railroad ties, sulfur, or general cargo; in 1901 the *Rebecca Palmer* took case oil to Europe – the first 5 master to cross the Atlantic.

1908 brought a slowdown, caused by a countrywide recession and specifically the increased use of steam driven colliers to compete with the sailing ships. And then, very unexpectedly, in September 1909, came the death of William Palmer at the age of only 50. His wife Marie ran the business herself for a year out of the Dorchester, Mass. office but then under pressure from some of the stockholders, sold the fleet to the J.S. Winslow Company of Portland. White ships became black.

Despite its short lifetime the Palmer fleet had been one of the big business successes of the early 20th century due in part to the owner's vigilance and penny pinching insistence. In another article we will examine Mr. Palmer's relationship with builder George L. Welt of Waldoboro.

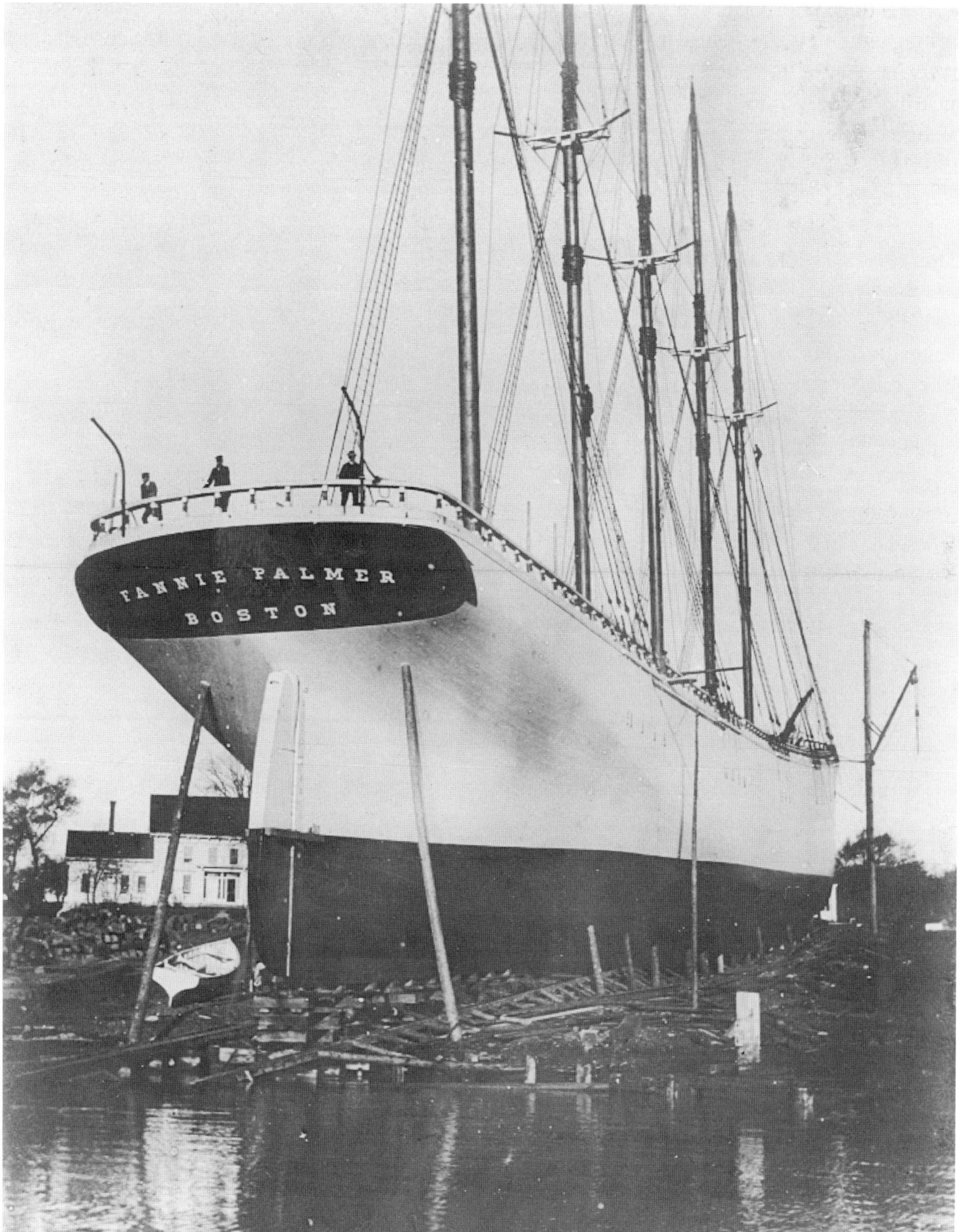

The Lincoln County News – September 20, 1900 – "1600 bushels of salt will be used on the schooner *Fannie Palmer* – placed between outer planks and ceiling (inner planks) to preserve the wood." – *Photo courtesy of Captains Douglas K. and Linda J. Lee*

Fannie Palmer – *Courtesy of Captains Douglas K. and Linda J. Lee*

The Palmer Fleet's Centennial - 1900-04
Fannie

2000 – 2004 marks the 100th anniversary of the famous Palmer fleet of giant freight carrying schooners built locally at the Town Landing site. It was on this spot that George L. Welt (nephew of long time ship builder Augustus Welt), financier William F. Palmer of Boston, and a 125 man crew began in early spring 1900 to "revive Waldoboro's great industry". From this spot the 6 enormous white vessels, largest ever seen on the river, would be launched – first the *Fannie Palmer*, then the *Baker Palmer,* then the *Paul Palmer*, the *Dorothy Palmer*, the *Singleton Palmer*, and finally in 1904 the *Harwood Palmer.*

Part of a 15 member wooden, wind driven schooner fleet (the others built at Bath, Rockland, and Boston), all painted white, all with a name ---- *Palmer*, these vessels were for general freight carrying but most often for coal. They were all managed by William F. Palmer, the school master turned businessman, whose tight fisted tactics made rich profits for the firm but often drove his employees (like Mr. Welt) to distraction. This grating personality has been discussed in an earlier story, and we will not repeat it here. Instead we will focus upon the 6 vessels themselves. And we begin with the *Fannie Palmer.*

Fannie Palmer, named for the owner's infant daughter, slid into the cold Medomak November 9, 1900. She was 266 feet long, 46 feet wide and 25 feet deep and was measured at 2075 tons carrying capacity. The tide that launched her was one of the highest in recorded history, bearing driftwood all the way to the Mill Street bridge. Towed by the tug *Seguin* of Bath, she started down river only to run into trouble. She was traveling light, and as a result high sided; a freshening east wind began to make her unmanageable, forcing her to slide sideways. Inexorably she drifted to starboard and ground out on the Western "Middle Grounds" - stuck – high and dry as that giant tide retreated. The day turned into night, and days turned into weeks as tugs struggled in vain to pull her off. Finally another tactic was tried – all able bodied men from the yard and many others were given shovels and fell to in an attempt to dig a canal from the *Fannie* to the main channel and float her off. Time and time again, the canal was completed only to fill back in with mud as the new tide entered. At last, after weeks of frustration the canal was successful: the schooner pulled back into the channel and towed without further trouble down the rapidly freezing river.

Commanded by George V. Williams, James E. Creighton (Palmer's favorite), Ernest L. Nash and others, the *Fannie Palmer* enjoyed an active and successful career, although her luck after launching was not the best: she went ashore twice, nearly burned once, and was in collision twice. In 1911 she became stranded near Cape Henry, Virginia and remained there, abandoned for several months. Finally salvaged and renamed *George R. Hudson,* she sailed again. On October 11 of the same year she caught fire at Norfolk, Virginia while loading coal for Boston; fortunately she was rescued and the fire extinguished by a crew from the tug *Virginia.* Then in October of 1913 she collided with the schooner *Margaret Haskell* of Vineyard Haven, Massachusetts escaping with minor damage.

Fannie Palmer's end came on July 11, 1914. Near Great Round Shoal, Nantucket, Massachusetts she was rammed and sunk by the steam collier *Middlesex*. Of the crew of 12, nine were rescued, two killed in the impact, and her master, Captain John Thomas, followed the tradition of the sea by going down with the ship.

Less than one year after *Fannie's* launching in Waldoboro, came the second of Palmer's Great White Fleet: *Baker Palmer* – October 1901. The town's hopes for a full scale revival of wooden shipbuilding were continuing.

Sources for this story included: *The Lincoln County News*, Stahl's *History of Old Broad Bay and Waldoboro,* the papers of long time customs agent Robert Applebee, and The *Waldoboro Bicentennial Book.*

The Lincoln County News – December 6, 1900 – "People are still talking about the high tide of November 9 when the Fannie Palmer grounded onto the flats." – *Photo courtesy of Captains Douglas K. and Linda J. Lee*

The Fannie Palmer.

Launch of Waldoboro's Greatest Production.

WITNESSED BY MANY PEOPLE.

A Most Pleasing Spectacle.

The staunch and beautiful five-masted schooner "Fannie Palmer" was successfully launched from the yard of George L. Welt, at 10.38 o'clock this morning.

This splendid specimen of marine architecture was built by our honored townsman George L. Welt for William F. Palmer of Boston.

The keel was laid last May and the work has progressed very rapidly under favoring skies.

The dimensions are as follows:

Length of keel, 257 feet.

Length over all, 300 feet.

Breadth of beam, 44 feet, 6 inches

... feet.

ceives the highest rating of her class.

She was built by George L. Welt of Waldoboro for William F. Palmer of Boston, who is the principal owner, having associated with him parties in Massachusetts, Rhode Island and New York and Capt. George V. Williams of Boston, who will command her. The schooner is designed for the general freighting business.

Capt. Waldermar of Boston, will go out as first mate with George Long as second mate.

Capt. George V. Williams, the proud commander of the Fannie ... has the record of a success- ... ...nded

ful master, having commanded vessels for twenty-one years, and sailed for Reed, Welt & Co. sixteen years. He has commanded the Maggie G. Hart and the Isaiah Hart, his last command being the four-master Marie Palmer, of Boston, built by Hon. William Rogers of Bath.

The schooner is named for the infant daughter of the principal owner.

George W. Young of Waldoboro, carried on the construction as master workman until the state of his health compelled him to give way to William A. Black of Bath. The several departments of work were bossed as follows:

George W. Purington, Fastner, Bath.

Reddington Cramer, Blacksmith, Waldoboro.

A. C. Erskine, Liner, Alna.

Standish & Ludwig, Caulkers, Waldoboro.

G. O. Waltz, In-board Joiner, Waldoboro.

Thomas P. Benner, Out-board Joiner, Waldoboro.

Osgood Miller, Painter, Waldoboro.

Wilford S. Mank, Spar-maker, Waldoboro.

S. A. Jones, Sailmaker, Waldoboro.

John Longley, Rigger, Rockland.

George H. Douglas, Yard Engineer, Waldoboro.

All the composition castings were furnished by Watson, Frye & Co., Bath, the blocks by J. S. Jackson & Son, Bath, the boats by J. B. Loring, Rockland.

... dropped into

Net tonnage, 2075.

Carrying capacity, 3700 tons.

The frame is of extra heavy selected white oak. The heavy deck beams are of hard pine strongly supported and braced with heavy knees. The keelson is 88 inches wide and seven feet deep, the measurement from top of keelson to bottom of keel being nearly eleven feet. The 70,000 feet of hardpine in the keelson is bolted together with 30 tons of iron.

Four large hatches 11 by 15 feet afford ample facilities for the rapid handling of cargo.

The masts are splendid specimens of Oregon pine, 112 feet long, 28 and 29 inches in diameter, surmounted by topmasts of the same, sixty feet long. The spanker boom of Oregon pine is 80 feet long, the other booms which are 45 feet long, and the gafts are spruce. The jibboom which is eighty feet long, is of Oregon pine. She will spread about ten thousand yards of canvas.

The Lincoln County News
November 1900

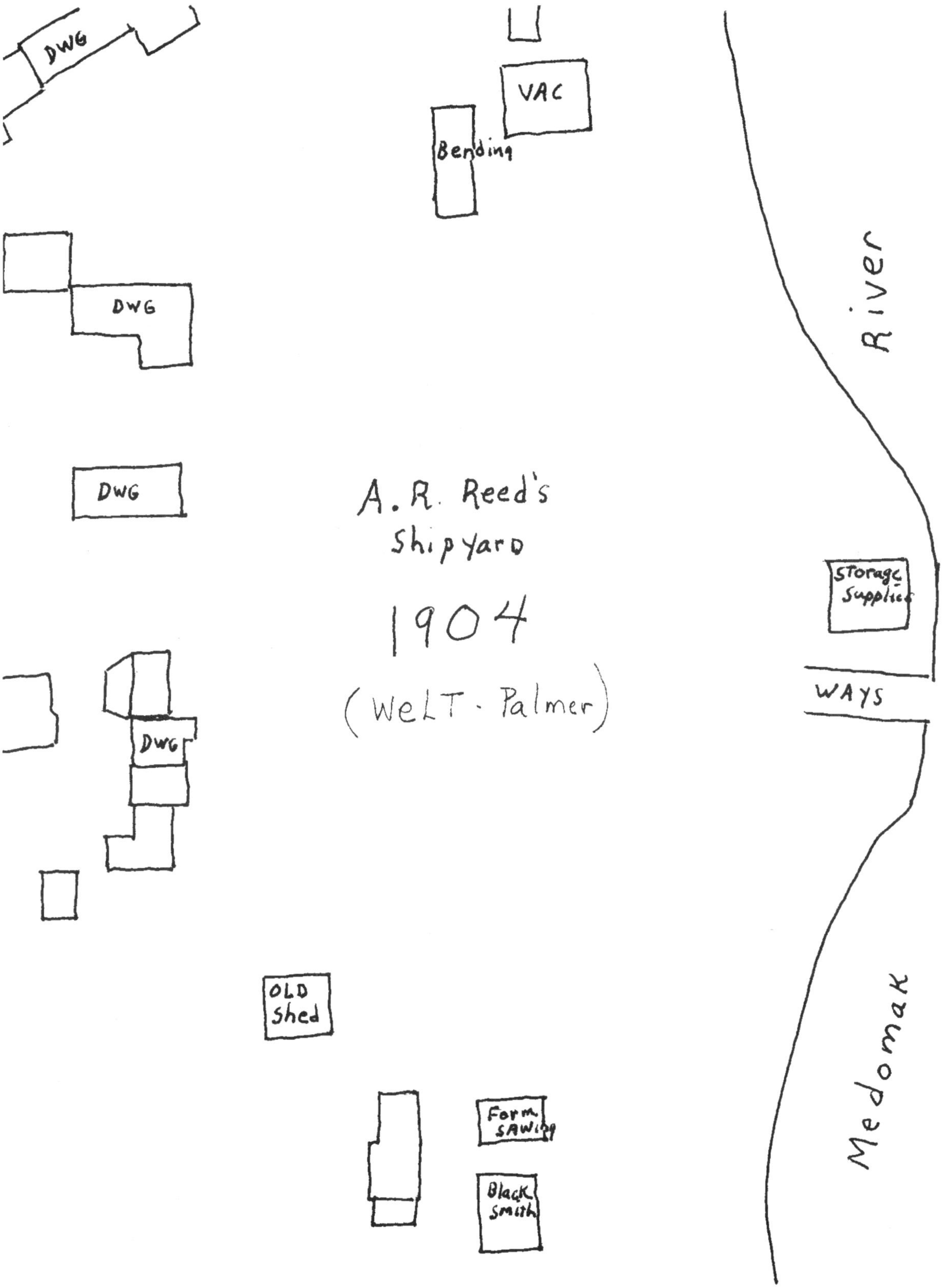

Map of Reed/Welt/Palmer shipyard. – Courtesy Waldoborough Historical Society

Photo courtesy of Captains Douglas K. and Linda J. Lee

The Lincoln County News – May 24, 1900 – "Mr. George Welt undoubtedly has the largest shipbuilding plant in New England – his shops, frame, timber, etc. cover 15 acres."

L.C.N. – June 21, 1900 – "Mr. Welt's yard embraces everything found in a modern shipyard. Over the engine room there is a cozy office with a telephone connection. There are now 100 men working, and this number will increase later."

Courtesy of Captains Douglas K. and Linda J. Lee

Baker Palmer – *Courtesy Maine Maritime Museum*

Baker Palmer – *Courtesy Ivan Flye Studios*

The Lincoln County News – August 29, 1901 – "The schooner *Baker Palmer* will be launched September 30. Harry Bohndell is putting the masts into the big schooner. A steam engine on board provides power."

L.C.N. – September 19, 1901 – "While hoisting one of the *Baker Palmer*'s big masts Saturday the lashings at the top of the shears slipped. Fortunately, the mast was only a few feet from the ground at the time, and the only damage was broken cross trees, easily repaired. The shears were taken down and relashed; work of masting resumed Tuesday."

The Palmer Fleet's Centennial – 1900 - 04
Baker

William F. Palmer launched his second giant coal carrying schooner (and the fifth for his fleet of 16) from the George L. Welt yard in Waldoboro, Tuesday, October 1st, 1901. As ever, *The Lincoln County News* reporter arrived to carry the story.

"Tuesday, October 1st, was a beautiful autumn day, and people began to come into town early by team and train. As the hour of launching approached, both banks of the Medomak were black with spectators – estimated at over 2,000. The majestic vessel, *Baker Palmer* (named after a benefactor, Captain L.D. Baker from the Boston area) was gaily dressed in bunting and flags of the signal code.

At 10 o'clock master mechanic John Peterson sent the crew of 25 men under the vessel to split out the blocks on which she had rested during construction. At exactly 11:25 she began to move down the ways while yet more than a dozen blocks remained. Then the "bridle" at the bow was cut and the beautiful craft slid almost noiselessly into the water. The anchors were dropped and the *Baker Palmer* was afloat in the Medomak. As she dropped off the ways, she was christened by Miss Martha Baker of Boston with a bouquet of native flowers. A lunch from the bakery of Schmidt, Miller & Co. was served to 100 guests on board soon after the launch."

The *Baker Palmer*, at the time the largest 5 masted schooner in existence at 284 feet long at the keel, 46 feet wide, and 26 feet deep, was begun in December of 1901 by builder Welt and his able bosses: Fastners – Frank Purington; Blacksmiths – Reddington Kramer; Caulkers – M.W. Standish & Sons; Inboard Joiners – G.O. Waltz; Outboard Joiners – Thomas P. Benner; Painters – Osgood Miller; Sparmakers – A.B. Sampson; Riggers – Henry Bohndell; Plumbers – C.H. Curtis; Yard Engineer – G.H. Douglas. Captain George Colby of Bath was to command the *Baker Palmer*; she was to be towed by the tugs *Knickerbocker* and *Stella* to the bay where her fittings were completed in a week. Her first assignment was to sail to Newport News, Va. and take on a load of coal for Boston.

Palmer's vessels were equipped with many modern conveniences. The cabin and other houses on *Baker* were heated by steam. The entire elegant furnishings of the aft cabin (finished in beautiful oak, mahogany, and cherry) were from Jordan Marsh & Co. of Boston. Two 25 Horse Power Hyde steam engines, fore and aft, a steam capstan and a highly geared steering gear made the work of the 12 man crew much easier.

The 5 masts of Oregon pine (Douglas fir) were 116 feet long - topmasts towered 60 feet higher. They measured 30 inches in diameter at the base. What a sight she must have been to William Palmer himself as he watched her sail past his house. But ever the controller and never quite satisfied, he could not resist the following comment in a letter to Mr. Welt, October 24, 1901:

"*Baker* was a beautiful sight as she sailed by our house. But her bow and bowsprit are too high. We must not have this again."

The *Baker* enjoyed a 14 year career as one of the best known fore and afters in the Atlantic coast coal trade employed by Mr. Palmer and, after his death in 1909, the J.S. Winslow Co. of Portland. Finally, in 1915, she was lost at sea off Bermuda, but her master, Captain Perry Denton, and his crew were rescued by the vessel *Rhine* and carried safely to Buenos Aires, Argentina.

Sources for this article include: *The Lincoln County News*, The Palmer letters – Maine Maritime Museum, and the writings of Robert Applebee.

THE BAKER PALMER

LARGEST FIVE MASTED SCHOONER AFLOAT

Successfully Launched from a Waldoboro Shipyard.

Tuesday, Oct. 1st, the day set for the launching of the schooner Baker Palmer, was a beautiful autumn day and people began to come into town early by team and train. As the hour of launching approached both banks of the Medomak were black with people. It is probable that fully two thousand witnessed the beautiful plunge of the majestic vessel, which was gaily dressed with bunting and flags of the signal code.

At 10 o'clock master mechanic John Peterson sent the crew of twenty-five men under the vessel to split out the blocks under the keel upon which the vessel had rested during the construction. At exactly 11.25 she began to move on the ways, while yet more than a dozen blocks remained. The "bridle" was then cut and the beautiful craft slid into the water almost noiselessly. The anchors were dropped and the Baker Palmer was afloat in the waters of the Medomak. Then it was seen that her draft was 10 feet 4 inches forward and 11 feet aft, a very desirable condition contemplated in her design.

As she dropped off the ways she was christened by. Miss Martha Baker of Boston, with a boquet of native flowers.

A lunch from the bakery of Schmidt, Miller & Co., was served on board to a hundred guests after the launch.

The Baker Palmer was designed by William F. Palmer for whom she was built by George L. Welt. The keel was laid last December. Her dimensions are as follows:

Length of keel,	262 feet.
Length over all,	305 feet.
Breadth of beam,	46 1-2 feet.
Depth of hold,	26 1-3 feet.
Gross tonnage,	2792
Net tonnage,	2241
Carrying capacity,	4,000 tons.

The frame consumed over 530 tons of white oak and there are 80,000 feet of hard pine in the keelsons. She has three full decks with four hatchways 11x13 feet.

The five masts are of Oregon pine 116 feet long, one of which is 30 inches in diameter, the others 29 inches. The topmasts, also of Oregon pine are 60 feet long.

The sails, which required 10,000 yards of canvas, were made by C. E. Dennett of Bath.

The cabin is similar to that of the Fannie Palmer and affords as fine accommodations as any merchant ship afloat, being finished in oak, mahogany and cherry, showing the artistic skill of our townsman, Mr. G. O. Waltz.

The cabin and other houses are heated by steam. The entire furnishings and fittings for the after cabin are from Jordan Marsh & Co., Boston.

The Baker Palmer is provided with two 25 H. P. Hyde Link Motion Steam engines, one forward and one aft, Hyde's largest size patent stearing gear, Hyde's double-acting windlass and steam capstans, one Warren Steam Duplex Wrecking Pump, 7x8¼x6, one Hyde Duplex Wrecking Pump, 6x5¾x6, one Warren Duplex Circulation Pump 5¼x3¾x5, one Warren Duplex Boiler Feed, Pump 4½x2¾x4.

She is also provided with three iron tanks with capacity of 7,600 gallons and carries two 5,000 pound anchors with 210 fathoms of 2¼ tested chains, also 1200 pound stream anchor. These anchors were furnished by the Camden Anchor—Rockland Machine Company.

The construction of the Baker Palmer began under the superintendence of ... Bath, who was succeeded by John Peterson also of Bath. The several departments were under the following bosses:

Fastner, Frank Purington.
Blacksmith, Reddington Cramer.
Caulkers, M. W. Standish & Sons.
Inboard Joiner, G. O. Waltz.
Outboard Joiner, Thomas P. Benner.
Painter, Osgood Miller.
Sparmaker, A. B. Sampson.
Rigger, Henry Bohndell.
Plumber, C. H. Curtis.
Yard Engineer, G. H. Douglas.

Capt. George Colby, of Bath, was in charge on board of the schooner. On the same tide the tugs Knickerbocker and Stella towed her to the "Bay" where her fittings will be completed in about a week, when she will sail for Newport News, having chartered to load coal for Boston.

The Baker Palmer is owned by William F. Palmer of Boston and others of Boston and Capt. Thomas McLaughlin of Westport, Me., who will command her. Boston is the hailing port.

She was built under special survey of American Shipmasters' Association and receives the highest rating of her class.

The name Baker Palmer is composite, being in honor of Capt. L. D. Baker of Boston, and the builder.

The Baker Palmer has the distinction, not only of being the largest vessel built in Waldoboro, but the largest five-masted schooner afloat, being larger in gross tonnage than the Wm. C. Carnegie. She is 536 gross tons larger than the Fannie Palmer built here last year, and 1014 gross tons larger than the Governor Ames built here in 1888.

This is the second vessel built by Mr. George L. Welt for Mr. Palmer and the keel of the third, to be known as No. 7, will be laid this week. We trust this is good business for Mr. Welt; we know it is good for Waldoboro.

The Lincoln County News, October 1901

An excerpt from a news item that appeared in the county newspaper

Launching at Welt's Shipyard a Gala Event

The people began to come into town early by team and train. As the hour of launching approached, both banks of the Medomak were black with people. It is probable that fully two thousand witnessed the beautiful plunge of the majestic vessel, which was gaily dressed with bunting and flags of the signal code. As she dropped off the ways she was christened by Miss Martha Baker of Boston with a bouquet of native flowers. A lunch from the bakery of Schmidt, Miller & Company was served on board to a hundred guests after the launching.

A ship launching in Maine in 1901

was a social event as well as a news item for the shipping page

THE BAKER PALMER

Dimensions 284.9 x 46.5 x 21.9 feet—Tonnage 2241 net, 2792 gross

Pride ran high in the little Maine town of Waldoboro on that October day in 1901 when George L. Welt sent the BAKER PALMER into the narrow Medomak River. She was the largest five-masted schooner afloat.

The BAKER PALMER was a three-decker with four hatches. Her lower masts measuring 116 feet and topmasts 60 feet. All were of Oregon pine. The houses were all steam heated. The after cabins finished in oak, mahogany and cherry, were masterpieces of the ship joiners art. The equipment included two 25-horsepower HYDE link motion steam engines, one forward and one aft; HYDE's largest size patent steering gear; HYDE's double-acting windlass and steam capstans; and a HYDE duplex wrecking pump.

The BAKER PALMER was built for William F. Palmer of Boston, and was last owned by J. S. Winslow & Company of Portland, Maine. She was one of the best known fore-and-afters in the Atlantic Coast coal trade. Lost at sea in 1915, the entire crew of the abandoned schooner was picked up by the ship Rhine out of Boston and safely landed at Buenos Aires.

HYDE
DECK MACHINERY

HYDE WINDLASS COMPANY
BATH, MAINE

EAST COAST REPRESENTATIVE
Smith-Meeker Engineering Company, 157 Chambers St., New York 7, N. Y.

WEST COAST REPRESENTATIVE

Hyde Advertising pages courtesy of Mr. Michael Footer

Mr. Palmer and His Fleet (Part 2)
Mr. Palmer and Mr. Welt

George L. Welt was an enterprising fellow, the nephew of a famous Waldoboro ship builder, Augustus Welt. As a young man, he played baseball with the town team up on Sproul's Level; he was a member of the Village Cornet Band. Reports from present day older residents describe him as well liked and easy to work for. Waldoboro's life blood, the ship building business, was dying by the turn of the 19th century, and George sought to revive it.

William F. Palmer was always looking for a better deal when it came to the construction of his fleet of large, wooden, white, coal bearing schooners. In September of 1899 he sat down with George Welt in his Boston office and hammered out a contract to build at least one vessel. What exciting news this was for Waldoboro! Welt was to expand the old Reed Yard (at the present Town Landing) to cover 15 acres and would hire 125 – mostly – local men. With shops and storage facilities it would become the largest ship building plant in New England. In the fall of 1899 supplies began to arrive for the project and were reported by the alert *Lincoln County News*.

> "32,000 Blue Ridge Mountain treenails by rail"
> "Large order of juniper knees by rail"
> "One and one half million board feet of hard pine by rail and sea"
> "140 tons of iron by sea – and more to come by rail"

Work commenced in the spring of 1900 and continued into the summer. The first big white schooner, *Fannie Palmer*, was launched in November; Mr. Palmer was so satisfied with the job that he contracted with Welt for a second vessel, *Baker Palmer*, before *Fannie* was even in the water. *Baker Palmer* was complete by October, 1901 and was followed by *Paul Palmer* – August 1902, *Dorothy Palmer* – November 1903, *Singleton Palmer* – May 1904, and finally *Harwood Palmer* – November 1904. Every single Palmer vessel was larger than anything built on the River before (or since) with the *Harwood Palmer* the largest of all at 301 feet.

(As mentioned in the previous article, Palmer named his ships after either family members such as son Paul and daughter Fannie or by the last names of important stockholders in his coal carrying business such as Baker, Singleton, or Harwood. He owned 15 schooners in all – others were built at Bath, Rockland, and Boston.)

Also mentioned in the previous article was Mr. Palmer's meddling nature. Writing regularly 4 or 5 letters a day, he bombarded builders, captains, and shareholders with news, advice, and, in the case of the former two, demands. For the builder, like Mr. Welt, very seldom was anything quite good enough. His work was always questioned for shoddiness, punctuality, and cost. Veiled threats to pull out of Waldoboro were common, along with hints perhaps to build in STEEL elsewhere. Sometimes Mr. Palmer would play captain off against builder; sometimes he would play Welt off against another builder. Some examples from letters – Palmer to Welt – follow (from the Palmer Collection at the Maine Maritime Museum):

July 13, 1901 – "I have been reminded 3 times by certain well informed people, that the garboards of my Waldoboro vessel (*Fannie Palmer)* were not put on so they would stay. A treenail will never hold a 10" plank. I shall ask you to drive iron spikes."

July 29, 1901 – "Now really, can we expect an October 16 launching? Captain Baker and his family hope to be there, and it is best not to disappoint them. Disappointment always leads to unpleasantness."

August 6, 1901 – "People in Bath prophesy that you are going to have trouble with your keel blocks."

October 3, 1901 – "Pleased be advised that your measurers made a mistake of 10 feet in the length of the vessel. They said 284 feet – it should be 274. The Custom House in Boston has severe penalties for such flagrant errors."

October 24, 1901 – "*Baker* was a beautiful sight as she sailed by our house. But her bow and bowsprit are too high. We must not have this again."

October 31, 1901 – "Satisfactory builders are like satisfactory captains – hard to find. I would like to settle on one builder and stay while we build with wood. If you wish to be in a position to build for us again, be exceedingly careful in your delivery of the *Paul Palmer.*"

August 7, 1902 – "It would be most wholesome for you, a shipbuilder, to remember that in thousands of little things, a captain who has had the experiences of Captain Risley is your superior. We are apt to lose ourselves in our own point of view. The wise thing is to put ourselves in the position of other people."

Poor Mr. Welt. If he had only known that Palmer's criticism and humorless schoolmaster lectures were exactly the same for all his contracted shipbuilders – whether in Bath, Waldoboro, Rockland or Boston!

The final article on the Palmer fleet will offer more details about the vessels themselves.

(Information for this piece of writing came from the files of *The Lincoln County News*, the Maine Maritime Museum, Stahl's *History of Waldoboro*, and telephone conversations with Mrs. Marjorie Freeman, Mrs. Martha Boggs, and Mr. Stan Hanna.)

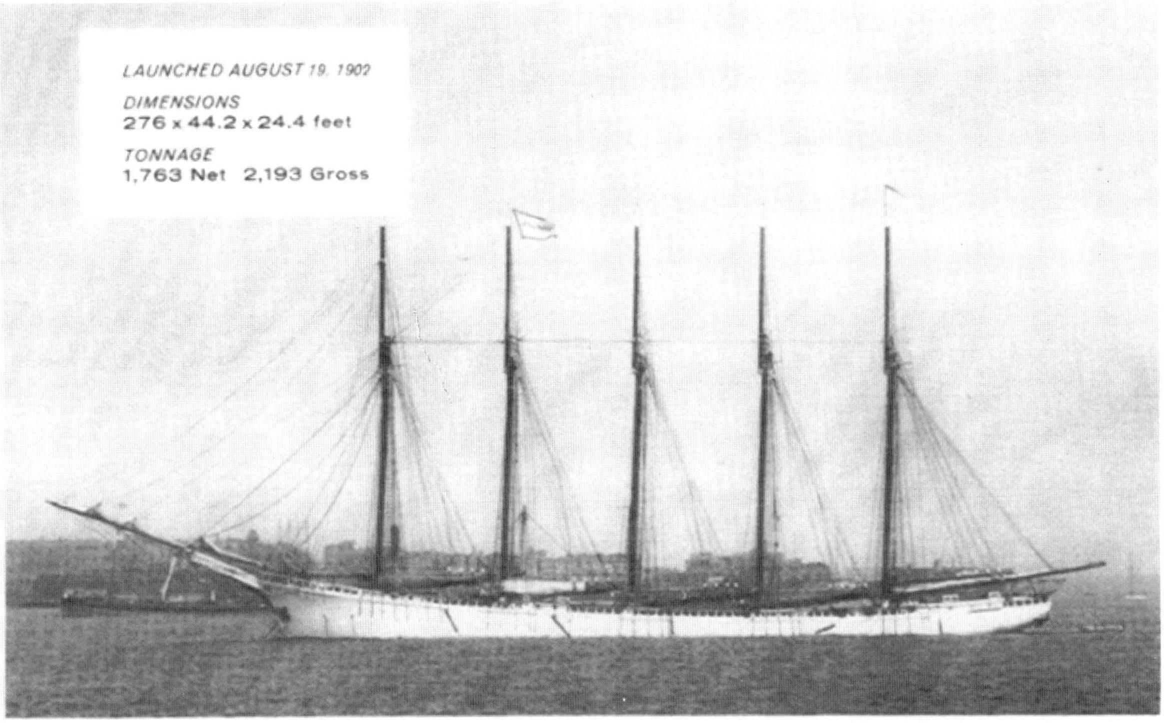

THE PAUL PALMER

HER LAUNCHING WAS OF COMMUNITY INTEREST

The Rockland, Me. Courier-Gazette commented, "At an early hour of the day, teams began to arrive from the adjoining towns, and every train brought large numbers . . . As she left the ways, Miss Florence Reynolds, daughter of Dr. Reynolds of Boston, gracefully performed the christening. The christening flowers came from the people of North Waldoboro and Waldoboro on the Washington Road. They were solicited and collected by Mrs. R. C. White and her daughter, Rose G."

The PAUL PALMER was the third schooner built by George L. Welt of Waldoboro for William F. Palmer of Boston. She was a beautiful vessel. No effort or expense was spared in the selection of materials and equipment.

The finest Virginia mountain oak, 450 tons of it were used for framing. Her planking required 675,000 feet of Georgia yellow pine. The Oregon pine for the masts and spars was brought to Bath by a sailing ship then rafted from Bath to Waldoboro. Her equipment included a HYDE reversible steam engine with a "modern" upright steel boiler, used to raise the sails and anchor and to load and unload. There were also two steam wrecking pumps, a HYDE windlass and patent steerer.

A fire aboard the schooner off Cape Cod in 1913 ended the career of this fine five master. Fortunately, a fishing vessel rescued the entire crew.

HYDE BUILDS—Steering Gears, Windlasses, Winches and Capstans for every size and class of ship in the Naval and Commercial Service.
Paper Mill, Textile, Centrifugal and Extruding Machinery, Machine Tools, Pulp Products Moulding Machines and other Special Machinery to Customers'

HYDE WINDLASS COMPANY
BATH, MAINE

WEST COAST REPRESENTATIVE

Courtesy of Mr. Michael Footer

COLN COUNTY

Terms: $1.50 a Year.
Single Copies Four Cents.

WALDOBORO, MAINE, THURSDAY, AUGUST 21, 1902.

LAUNCHING DAY.

The Paul Palmer Successfully Launched Tuesday.

Scenes and Incidents Attending the Event---Banquet on Board---Details of Construction-Minor Mention.

The Night Before the Launching.

Finished fair from stern to bow,
The great ship lay all shimmering white.
Tomorrow would bring the ocean s roar,
And she thrilled with a strange delight.

The moon looked from the sky above
"Oh, ship 'tis the last night of peace,
Tomorrow comes battle with wind and wave "
Cried the ship, "I call it release.

In this haven I hear but the sound of bells
And the sights of the village weary my eye,
But yonder for me lies the breath of life
In the boundless beauty of sea and sky.

The bird has flown to her nest long since
And folded her wings in rest,
And you in the sheltering arms of the sky
Shine jewel like on its breast.

But I—the waters cry and call,
This peaceful silence is not for me.
You to your place, and the bird to hers,
But I to the toss of the beckoning sea.

Oh, the outward plunge through the foamy
drift,
The strong, true arms, the hearts so brave;
Oh, the gathering murk, the fog bank's thrill,
And the tender clap of each white capped
wave;

For each of us has a life to live,
And some paths lie fair and fine
By the green of hills and flowering dales,
But destiny calls me afar to mine.

The ship to the sea and the moon to the sky
And the bird to its sheltered nest,
The same Heaven lies above us all
And we know what is, is best."

The stars grow pale in the rose flushed sky;
The first sweet breath of dawn steals clear
Over wakening land and expectant sea,
The day of joy, of release, is here.

A beautiful day was Tuesday, August 19, launching day, and every one, far and near were out in gala attire to do honor to the launching of the beautiful craft, which for

was turned to the point where Foreman Young, with his crew of some 20 men, were moving down to cut out the blocks under the keel, upon which the vessel had so long rested. One block after another was split out, and at last a quiver was seen to run from stern to stem. Everyone held their breath, so deep was the interest felt. One more block, and the vessel started on her way to the stream. At once the "bridle" was cut, and in a few seconds the Paul Palmer slid gracefully from the ways to the peaceful bosom of the Medomak. The anchors were

ed charters somewhat, still there is more or less coal moving all the time, and the owners of this line find no trouble in keeping their vessels well employed.

The crew will consist of Captain C. E. Risley of Pleasantville, N. J., first mate, W. S. Blake of Boston, second mate, cook, cabin boy and ten able seamen. The crew are from Boston and will join the vessel at Muscongus Bay, 3 miles below Waldoboro. Ten men would seem a small crew for so large a vessel, but steam displaces a good many and does the work quicker and surer.

This makes the third vessel to be built here for the firm of William Palmer. Previous to the year 1900, when the Fannie Palmer was built by Mr. Welt for this firm, they had had vessels constructed at Bath. It speaks well for Mr. Welt's business ability that he has been able to hold the custom of such a firm against the keen competition he has to encounter on every hand. The owners evidently know good work and appreciate the builder's efforts to furnish the same.

As soon as the yard can be put in proper shape the keel of another five master for the same parties will be laid. She is to be named Dorothy Palmer, and no expense is

The Lincoln County News

Paul Palmer *under construction – 1902 – Author's Collection*

Mr. Palmer and his Fleet (Part 3)
A Launching at Waldoboro

The launch of a Palmer vessel was an event one would not easily forget. Mr. Harold Castner, a historian from Damariscotta born in 1888, who attended several of them once wrote: "It was indeed a gala day. The great hulls were pure white, and with flags and bunting flying, and the host of people crowded into every vantage point to view the spectacle, the majestic beauties slid into water with extreme grace. They were christened by some female of the Palmer family or entourage and will long be remembered as the peak of perfection in shipbuilding grace and beauty."

For weeks before the launchings, Mr. George L. Welt, the builder, was very busy with the crowds who arrived to view the progress of his schooner. At almost every hour of the day visitors would be on the shipyard grounds, often mingling with the workers on and away from the vessel. (One photograph of the day shows Mr. Welt in his derby hat gallantly assisting a lady guest in disembarking.)

In the evening prior to the launching the customery "Launching Ball" was held at Clark's Hall on Main Street. For the *Paul Palmer,* third of the schooners, this event was August 18, 1902. The "Farewell Opera House Sextette" provided the music and 400 elegantly dressed people, many from out of town and guests of Mr. Palmer, crowded into Waldoboro's fine all purpose gathering place. *The Lincoln County News* reporter was also there and described: "At 9:00 Mr. George L. Welt himself and Miss Stella Murphee of Augusta led the Grand March and were followed by 100 couples. Many beautiful costumes – white and light colored fancy silk dresses for the ladies – added a picturesque appearance to the occasion as the ocuples went whirling through the mazes of the waltzes and quadrilles." Dinners were served the guests and participants at private homes or at the local hotels – the Savoy on Friendship Road and the Davis next door to Clark's.

The morning of the launch saw more interested people arriving by buggy, horseback, train, and boat (steam powered vessels as well as sail and rowboats brought visitors from as far away as Martin's Point in Friendship and Round Pond.) In the Welt shipyard the crowd steadily grew more excited. About an hour before high water, a cry of "Here they come!" signaled the approach of Edwin Young and his 20 burly launch crew members as they moved to their assigned places under the keel. With giant sledge hammers they would split out the great launching blocks beginning at the stern and moving up. When this job was complete, everyone held their breath; suddenly an enormous quiver could be heard running from bow to stern. And then slowly at first but then faster and faster the vessel would begin to slide, astride the wooden cradle down the steel tracks or ways. As she began to slide, a young lady would christen her with a bunch of Waldoboro grown flowers and the traditional prayer of "I christen thee the —— *Palmer*." As the launching crew sprang to safety, the huge white schooner would continue to slide into "the peaceful bosom of the Medomak" where her captain and crew would take over. She would be towed down river and proceed to her destination of loading coal (usually) for William L. Palmer and the homes and industries of New England.

The *Paul Palmer* sailed for the next six years without serious mishap. Then in July of 1908 she caught fire at a dock losing her topsails and foremast. She survived and went on until July 15, 1913. Off Race

Point, Cape Cod, she again caught fire and this time burned to the water's edge. Her captain at the time, Albert C. Allen, his wife, a woman guest, and the crew of 10 were rescued by the fishing schoooner, *Rose Doherty*.

Material for the article was procured from *The Lincoln County News,* the Rockland *Courier-Gazette*, an advertisement for the Hyde Windlass Company of Bath, and the records of historian Robert Applebee.

Launching Our Paul *– looking north – Courtesy Maine Maritime Museum*

Launching Our Paul *– looking west – Courtesy Maine Maritime Museum*

The *Dorothy Palmer* – The Longest Lived

On Thursday, June 4, 1903 at 11:30 A.M. the *Dorothy Palmer*, fourth Waldoboro Palmer vessel – eighth Palmer vessel in all, slid into the Medomak River from George L. Welt's yard (today's Town Landing.)

Painted white like all of her sisters, the *Dorothy* was, at the time, the largest five masted vessel afloat and was exceeded in size only by the six masters *George W. Wells* (Camden) and *Eleanor Percy* (Bath). She was 294 feet long, 46 feet wide, and 26 feet deep with a carrying capacity of 2315 tons. Her five masts of Oregon pine measured 118 feet with topmasts of 60 feet. Three of the masts were 32 inches in diameter at the base; two measured 30 inches. Her outer planking's thickness was five and one half inches. She was framed in Virginia white oak and strapped diagonally in iron.

At Waldoboro that June day 5000 people had gathered, arriving by boat, train, buggy and foot. Mr. Palmer, as was his wont, had brought in a large contingent of family and friends from the Boston area and celebrated with them and locals with lavish dinners and breakfasts as well as lodgings at one of the hotels in town. Clark's Hall on Main Street was swinging on the evening before with music and dancing – the gala Launching Ball. And after the successful, thrilling launching, many were invited to visit the ship as she lay, adorned with bunting, at anchor in the harbor.

The next day, Friday, "Our Dorothy", as Mr. Palmer liked to refer to his ships, was towed by the tug *Amelia* down river to Muscongus Bay from where she had been scheduled to sail to Hampton Rhodes, Virginia under the command of her skipper, J.C. Harding of Chatham, Massachusetts.

"Our Dorothy" was lucky enough to enjoy a remarkably accident free career – one which lasted twenty years, by far the longest of the Palmer vessels – and exceedingly long for the coal carrying sailing ships. (Indeed she was to outlive Mr. Palmer himself by some eleven years and the builder, George Welt, by three.) Her worst mishap occurred in 1912; while sailing light from Portsmouth, New Hampshire to Norfolk, Virginia, she was blown off shore and was missing for a month before finally being rescued and towed in by a revenue cutter.

She ended her long life on March 28, 1923, when she was driven off course by a northwest gale and onto Stone Horse Shoals in Nantucket Sound. Captain Willis Hutchins and his crew of ten were rescued the following day by a surf boat from the Monomoy Point Life Saving Station.

From the laying of her keel back in September of 1902 to her launch in June of 1903 and to her remarkable career, the *Dorothy Palmer* was on the sea later in the 20th century than any other Waldoboro vessel.

Information for this article came from the files of Robert Applebee, Marine historian, and also from a speech by J. Malcolm Barter, Waldoboro resident.

Schr. *Dorothy Palmer*

It is doubtful if anything short of three-ringed circus could attract as much attention in Waldoboro and vicinity as the launching of the mammoth five-masted schooner *Dorothy Palmer*, which takes place from George L. Welt's yard Thursday forenoon. The event has been looked forward to for many weeks and will be observed as a gala day in Waldoboro. *The Lincoln County News* will go to press half a day earlier in order that the whole force may share in the general rejoicing; the stores will close; a public dinner will be served; and in the evening the famous Rambler Club will give a ball. In order to make this latter function possible, a theatrical troupe consented to cancel its date. Trains will carry spectators from points along the Knox & Lincoln Division, while a special from Boston will bring to Waldoboro the Massachusetts people who are concerned.

The special features in this launching lie in the fact that it is the largest craft of any rig ever built at Waldoboro, while the incline on which the great craft sits, and the narrowness of the river which is to receive it, promises excitement as well as the beauty which always pertains to a launching. Not long ago a vessel which was launched in this river stuck fast on one of the muddy banks. The launching took place during a high course of tides, and as the water sank lower and lower each day it left the new schooner so far inland as to cause wonder how she came there. The vessel was extricated at the next high course of tides, after much difficulty.

The *Dorothy Palmer* was built for William F. Palmer and others of Boston, and her estimated cost is in the neighborhood of $130,000. Her dimensions are as follows: Length of keel, 273 feet; breadth of beam, 46 feet, 6 inches; depth of hold, 26 feet, 6 inches. The tonnage has not been figured.

The frame is of native oak, and the planking and ceiling are of hard pine. The builders say that the iron used in strapping the frame would equal a distance of 2½ miles if spread out. The schooner has three large houses, all elaborately finished, and provided with every convenience known to modern seafaring. Two Hyde engines of reversible type and 25 horse-power are included in the equipment. The vessel will be controlled by screw steering gear.

The masts are of Oregon pine and each 118 feet in length. The topmasts will measure 60 feet in length. The foremast is 32 inches in diameter, the mizzenmast 30 inches, and the other three masts 29 inches. The vessel will spread about 12,000 yards of canvas. She will be ready for sea when launched. The two principal anchors are of the stockless design, each weighing 9000 pounds. With them are 220 fathoms of chain.

George L. Welt, owner of the shipyard was master-builder, and George W. Young was master carpenter.

The frame for another five-master, to be known as the *Singleton Palmer* is already in the yard, and she will be immediately followed by another large schooner for the Palmer fleet.

A very large crew is constantly employed in the Welt yard, and with the granite quarries in full boom the town is enjoying a season of marked prosperity.

The Rockland Courier-Gazette, Tues., May 26, 2003

Courtesy of Captains Douglas K. and Linda J. Lee

Dorothy Palmer launching – *November 28, 1903 – Courtesy Maine Maritime Museum*

Dorothy Palmer – *Courtesy Maine Maritime Museum*

Singleton – One Hundred Years Ago

The winter of 1903 – 1904 was a frigid one. *The Lincoln County News* reported on February 11 that "no one has had the courage to venture out into the -20 degree weather to try sleighing". For the first time (with one exception) in thirty years the river froze so completely that teams could travel on the ice from the mainland to Bremen Long Island. During more than forty days, horses, wagons, and drivers were seen moving across Broad Bay and through the "Narrows" on their way to Bremen. (No doubt disappointing to children was also the news that, despite the cold, attendance at the district schools had been "excellent!")

Mr. William F. Palmer's fifth schooner's keel had been laid in June of '03. She was to be named *"Singleton Palmer"* after shareholders James and Ernest Singleton, worsted wool manufacturers of Woonsocket, Rhode Island. Her length was to be 269 feet; width 45 feet; depth 21 feet; tonnage 2859. The builder as usual was George L. Welt; master foreman was George W. Young; rigger Henri Bohndell; caulker Miles Standish among other leaders. All through the summer, fall, and winter the 100-125 man crew labored.

The cold winter had caused several interruptions in the schedule and raised havoc with Mr. Palmer's famous lack of patience. He had been unhappy with Mr. Welt for some time, badgering him mercilessly and continuously through the mail about the quality and punctuality of his work. In the early spring behind the scenes he gradually brought an end to Mr. Welt's title, replacing him as boss with an old favorite – Captain Jim Creighton of Thomaston. Creighton had been skipper of the *Fannie Palmer* and was preparing himself to command *Singleton* when the word came for him to move in as boss of the shipyard. It was announced that A.V. Potter was to be the captain instead of Creighton; Creighton's replacing Welt was not formalized. But it was real – much to the dismay of Welt who found himself odd man out. Nothing much could be done – and the building had to go on. Welt protested; Palmer owned the business.

On launch day April 16, 1904, a heavy snow fell – one half a foot on the level and piling up everywhere. Crowds of enthusiasts, having enjoyed the traditional "Launching Ball" at Clark's Hall the night before, were held back by the weather, and places such as the Hotel Savoy on Friendship Street which had promised a meal to anyone who wished it, suffered by the lack of attendance. The launching procedure itself was held up 20 minutes to frozen equipment, but at 10:50 the enormous white vessel slid into the Medomak. Unfortunately she slid too far – shooting over to the east side, ramming into the bank and damaging the rudder and stern post. Aground near Storer's Wharf, the *Singleton Palmer* remained while a cofferdam was constructed and the sternpost repaired. One and one half weeks later she was towed by two steam tugs down river to deeper water where the new rudder could be installed. On the 12th of May the *News* reported that the new rudder had been attached with no difficulty, and the *Singleton* was on her way to Newport News, Virginia for her first assignment in the coal carrying trade.

Like her Palmer sisters, the *Singleton Palmer* never again saw Waldoboro. She spent a long career on the sea and finally came to an end off F9enwick Island, Delaware in 1921 when she was rammed and sunk by the Clyde liner *Apache*. Captain J.M. Griffin and eight of his men were rescued by the steamer *Gloucester* and taken to Boston, while the mate and another man managed to reach the *Apache*. One crew member was lost.

Sources for this article included *The Lincoln County News*, an advertisement for the Hyde Windlass Co. of Bath, Maine supplied by Mr. Michael Footer, the writings of historian Robert Applebee and a conversation with Captains Douglas K. Lee and Linda J. Lee.

Captain Jim Creighton
Courtesy Maine Maritime Museum

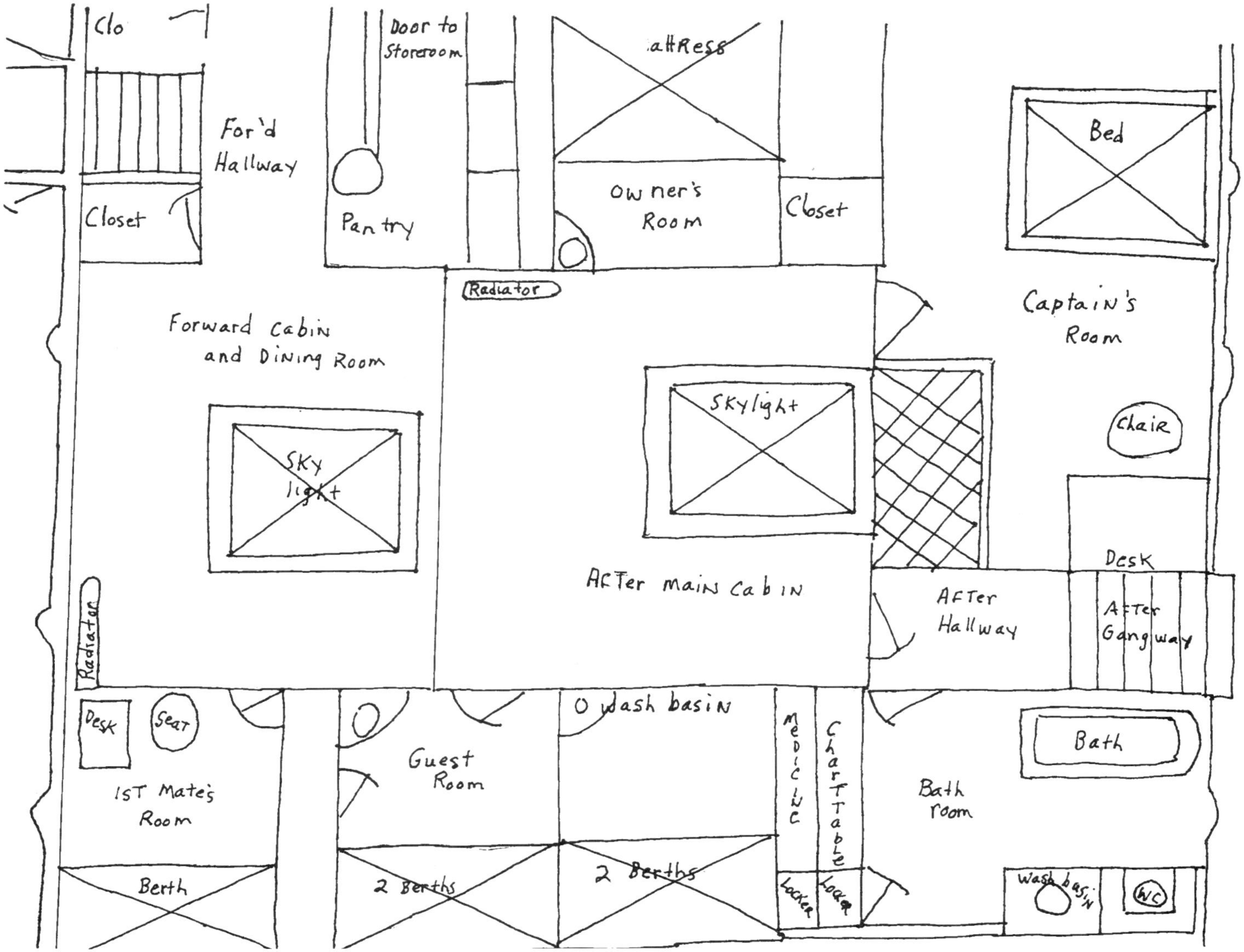

Aft quarters of the Jane Palmer, *built in Boston in 1904 as part of the fleet. – Courtesy Maine Maritime Museum*

Singleton Palmer – *Courtesy Mr. James Stevens*

The Singleton Palmer

Owner: W. F. Palmer of Boston
Built by: Geo. L. Welt at Waldoboro, Me.
Launched: April 16, 1904

A snowstorm at her launching caused financial hardship in the town of Waldoboro

The Lincoln County News of nearby Damariscotta reported:
"The storm made a great difference to the town financially, for had it been a nice pleasant day there would have been quite a business at our stores, while as it was there was nothing. They were not the only ones to suffer. The ladies of the Baptist and Methodist societies had made preparations to feed large numbers at dinner and as it was they were obliged to abandon it entirely. Then the Hotel Savoy and the Bakery had also made great preparations for large numbers and they, too, were doomed to disappointment and loss. But that night, at least, there was a proper dance in honor of the launching; it was well attended despite the inclement weather, about 60 couples participating, and a very pleasant evening was passed."

Built from a 269-foot keel, the five masted schooner Singleton Palmer was the fifth big fore-and-after built at Waldoboro for W. F. Palmer's fleet out of Boston.

The schooner, completely HYDE equipped, had a Virginia oak frame strengthened by 160 iron straps. She was a three decker spreading 11,000 yards of canvas. Built for the coal trade, the Singleton Palmer had a capacity of 4,400 tons.

A collision on November 6, 1921 with the S. S. Apache off Fenwick Island Lightship brought an end to the big schooner. At the time of her loss the Singleton Palmer was owned by the France & Canada S. S. Corporation and was valued at $45,000.

HYDE BUILDS—Steering Gears, Windlasses, Winches and Capstans for every size and class of ship in the Naval and Commercial Service.

Paper Mill, Textile, Centrifugal and Extruding Machinery, Machine Tools, Pulp Products Moulding Machines and

HYDE WINDLASS COMPANY
BATH, MAINE

East Coast Representative
Smith-Meeker Engineering Co., 157 Chambers St., New York 7, N. Y.

Courtesy of Mr. Michael Footer

LINER AND SCHOONER CRASH, ONE MAN LOST

11 of 5-Master's Crew Rescued after Head-On Collision Sixty Miles Out.

SEEK SAFETY IN RIGGING

Singleton Palmer's Deck Engineer Missing Following Early Morning Accident in Calm Sea.

The Clyde Line steamship Apache, bringing eighty passengers, arrived at her pier, 27 North River, six hours late last night after having collided at sea yesterday with the five-masted American schooner Singleton Palmer, which was left lying on her beam ends about sixty miles southeast of Scotland Lightship near Fenwick Island. The deck engineer of the schooner, [illegible] Swanson, was missing after the collision. The remaining eleven members of the crew were rescued by boats from the Apache and the Merchant and Miners Steamship Company's liner Gloucester.

The bow of the Clyde liner was damaged, portions of the schooner's bulwarks still sticking to its last night. Blankets and pillows were used to stop the flow of water into the ship forward. The Apache will probably have to go into dry dock.

The sea was calm and the wind was fair when the collision occurred. Clyde Line officials were reticent concerning the cause of the accident and would not permit reporters to question the Apache's officers last night.

Members of the crew said the schooner crashed into the steamship head-on early in the morning. The Apache promptly lowered a lifeboat and rescued First Mate Sven Landstrom and another member of the crew, while the Gloucester, passing nearby, picked up Captain Griffin and eight other members of the schooner's crew. Thomas Jones, an able seaman, climbed the forward rigging after the crash and managed to reach the deck of the Apache while her bow was still interlocked with that of the schooner.

The schooner was bound from Perth Amboy to Norfolk in ballast. The Gloucester was bound from Norfolk to Boston, where she will land the nine survivors she rescued.

The Singleton Palmer was a five-masted, double-decked schooner of 2,839 gross tonnage, 294 feet long and of 45-foot beam, built at Waldoboro, Me., in 1904 and one of the few five-masted American schooners remaining in the coastwise trade.

The Apache was in charge of Captain W. O. Chichester and carried a general cargo from Jacksonville, Fla., and Charleston, S. C., in addition to passengers. The Clyde Line has had other bad luck lately. Another of its steamships in the Florida trade was in a collision recently, and the latest addition to its fleet, the Lenape, sank at her pier a short time ago and then capsized after having been raised

Courtesy of Captains Douglas K. and Linda J. Lee

The Final Vessel - *Harwood Palmer*
Part one - William Palmer vs. George Welt

The question of whether Mr. William F. Palmer of Boston would build yet another vessel in Waldoboro was answered in late March of 1904. A skeleton or frame arrived by railroad in pieces and was hauled from the depot to the Welt yard. There it rested while the building crew painted, rigged, and readied the *Singleton Palmer* for launching April 16. By May 5 steamer *Seguin* had towed *Singleton* down river, and the shipyard, never ceasing activity, saw the new frame set up for the sixth and final member of the fleet.

As far back as the end of 1902, Mr. Palmer had been unhappy with master builder George L. Welt, the man who had been responsible for bringing him and employment for 125 workmen to Waldoboro. Writing letters several times a week, he critiqued Welt's work, complaining about quality, punctuality, and cost. Over and over he would threaten to move the plant to another location, build with steel, and question the choices of materials and men. By the fall of 1903 he had decided to ease Welt from his responsibilities; calling a meeting in Waldoboro he publicly pointed out debts owned to workers and investors and announced that Captain James V. Creighton would serve as interim superintendent of the yard. For the first time, employee Welt protested publicly against employer Palmer. In a long article published in the Rockland *Courier - Gazette* September 26, he stated that he 1) did owe money but not *that* much and that it would be all paid back after the next vessel was completed; 2) Captain Creighton was *not* taking over for him – he was merely around in his capacity as future master checking on progress, "a usual proceeding". He also reminded all that it was he who had brought shipbuilding jobs and hope back to Waldoboro after a number of years of "grass growing amok in the yards". And he asserted that he had been offered a better job elsewhere as a superintendent and would probably be taking it where his accomplishments over the last 25 years would be more appreciated.

Mr. Palmer, through his Waldoboro attorney William H. Miller, answered back the following week with his own story also published in the *Courier - Gazette*. Much more money was owed by Mr. Welt, he said, than had been admitted to – probably close to $80,000 – not only to employees but also to businesses and banks. Mr. Welt's personal credit was VERY suspect; he was not to be trusted any more, and as far as responsibility for the new vessel, Captain Creighton was to be totally in charge, while "George Welt might just as well go a – fishing until the schooner is launched and ready for sea."

Probably because of Welt's respected family name in Waldoboro, he was kept on the yard roster through the summer and fall of 1904, at a salary of $18.00 a week, as the building went on. He was left hanging by Palmer – never being really leveled with, expected to "keep an eye on the ship's progress and report" it to Palmer who stayed in Boston. All this he did – writing to Palmer and traveling by train to visit with him, hoping always for his old job back. Meanwhile Palmer continued his correspondence with Captain Creighton through the summer, occasionally referring to Welt as a nuisance factor to be tolerated – for the time being. We have these letters – thanks to Captains Douglas and Linda Lee of Rockland. Some quotes follow:

May 26: "Geo. Welt is here and we flitted away all day yesterday without our being able to say a word about the matter at hand. He is stuttering pretty badly but says you are getting along very slowly on the vessel. That you are losing time all the way on the schedule. The very best season is going and we shall be in an awful plight with our investors if we don't get that vessel out of Waldoboro soon after she is launched."

May 31: "Geo. Welt ran off to Waldoboro before I got through with him and without my permission, but he was so drunk when he went that I shall have to excuse him. Such a reasonable man as you should not have to waste so much time and ink on this fellow."

June 2: "Continue to be as civil as possible to Geo. Welt. I have tried to convey to him that you do not want him too much under foot, and that I wanted him to live in peace with you lest he lose his $18.00 per week that you were protesting about on the payroll. I now give my attention to a dozen other troubles."

August 31: "The *Hon.* Geo. L. Welt writes me this morning. I had previously written him that if he wanted to see me, he would have to come to Boston, but, from now on, at his own expense."

And finally – on October 12 – one month before the launching of the 6th Palmer schooner, *Harwood Palmer*: "Mr. Welt is with me this morning and is listening to this clause in the dictation of your letter: 'Please be advised and directed to take from the payroll the name of George L. Welt and to see that that name is not again extended on the yard account of this vessel.'"

The agony was at last over.

Sources for this article included *The Lincoln County News*, the *Rockland Courier Gazette*, and the Letters of William F. Palmer to Captain James Creighton collected and bound by Captains Douglas and Linda Lee.

The Final Vessel – *Harwood Palmer*
Part Two – End of "The Great Days"

William F. Palmer of Boston, owner of five enormous coal carrying white wooden schooners built in Waldoboro between 1900 and 1904, had solved his acrimonious falling out with his builder, George L. Welt, by firing him in October 1904. Welt's replacement, James V. Creighton of Thomaston, received letters nearly every day from his employer (as had all builders who had ever tried to work for and please Mr. Palmer). In his usual nit-picking manner, Palmer pushed hard to finish the new vessel, *Harwood Palmer*, before winter set in. "I have known Waldoboro full well enough in icy weather. There is only one thing to do: Take out the whip and drive to the wire – you will have to tell the Waldoborians that the cause warrants working even some Sundays." At the same time he took time to praise some of the "the Waldoborians" for good work and compare them most favorably with some of the men from his Boston yard also working on a schooner. "Why can't I bring some good caulkers and their boss, Myles Standish, down here to work with this bunch?"

By fall of 1904 Mr. Palmer was also becoming weary of local conditions out of his control besides the weather: a narrow launching site, treacherous tides, and a shallow river. He strongly urged Creighton to move the vessel out to sea as quickly as possible after launching. "Every indication is for an early winter. The hair of all animals has grown as long as it usually does in January! Do not let us fail now that we have got so near the end."

And unlike previous Palmer optimism, he made no promises whatsoever to Waldoboro folks concerning future vessels to be constructed.

The *Harwood Palmer*, named for a Boston benefactor, was the largest of all the Palmer fleet members at 301 X 46 X 27 feet and would be able to carry 2885 tons of cargo. She slid on National Election Day (Republican Theodore Roosevelt defeated Democrat Alton Parker and Socialist Eugene V. Debs), November 8, 1904, into the cold Medomak and enjoyed a 13 year career as a coal carrier. The fortunes of wind driven wooden sailing ships were declining now because of steel, steam, and railroads; but the advent of World War I brought along a short lived "Indian summer". Coal rates soared up to $20.00 per ton; voyages to the blockaded states of Europe were open for any vessel whose owners dared take chances with German U-Boats and mines. The *Harwood Palmer* had survived one horrific collision with the steam tug *Juanita* back in 1909 when both vessels in thick fog lost sight of one another off Highland Light, Cape Cod. According to historian Robert Applebee, the bowsprit of the *Harwood Palmer* ripped straight through an unoccupied stateroom of the *Juanita*, so completely damaging the sailing vessel's headgear that she had to be towed to Boston for repairs.

On May 23, 1917, however, the *Harwood Palmer* was not so fortunate. Near the town of Nazaire, just off the French coast, she struck a floating German mine; the ensuing explosion sank the *Palmer* quickly. Happily, all members of the crew were able to row to shore in safety due to a plentiful number of small boats carried on board the big schooner.

Neither William F. Palmer nor George L. Welt were present to be told about the demise of their last vessel. In September of 1909 Palmer had died from acute rheumatism at the relatively young age of 50. His fleet – which had topped out at 15 white schooners – all named --- *Palmer* – was passed on to his wife, Marie, and her advisors. After a year of attempting to operate out of the Dorchester, Massachusetts office, Marie sold the business to the J.S. Winslow Company of Portland, Maine. George Welt, Palmer's unfortunate cohort, also had passed away – in 1917 – at the age of 62. Waldoboro's shipbuilding days were over.

(Sources for this article included *The Lincoln County News,* the Letters of William F. Palmer to Captain James Creighton collected and bound by Captains Douglas and Linda Lee, the writings of U.S. Customs House Director and marine historian Robert Applebee, and *A Shipyard in Maine* by Ralph L. Snow and Captain Douglas K. Lee.)

Harwood Palmer *(named after Boston area benefactor) Launched November 8, 1904 – 301 feet x 46 X 27 feet; 2400 tons – Courtesy Maine Maritime Museum*

Fate – Her building was finished by James E. Creighton of Thomaston, also her first commander. In her first year of service she went aground off North Dennis, Cape Cod, and it took 5 months to dig a channel across the mud flats to free her; then she almost immediately collided with the steamer Juanita *off Nantucket in fog. In 1914 she barely escaped a gale by being able to slip into Provincetown harbor. Finally, her luck ran out in 1917 when she was attacked and sunk by a Geman submarine near Nazaire, France.*

Dimensions: 301.7 x 46.3 x 27.8 feet

Seldom is the interest of a town divided between a presidential election and a ship launching, but such was the case of the town of Waldoboro, Maine, on Nov. 8, 1904. On that date the schooner HARWOOD PALMER slid from the ways at Welt's Shipyard. The result of these competing events was best described in a contemporary newspaper which noted: "Up to this time the Town House, the place of voting, had been entirely forgotten, but after the launching many went to vote before going home. * * The bakery as well as the Baptist and Methodist societies were kept busy until late in the afternoon feeding those from out of town."

The big five-masted schooner carried 11,000 yards of canvas. She was framed with Virginia oak, reinforced with 160 iron straps running at an angle of 45 degrees from belt strap to short floor heads. Where they crossed at intervals of four feet, the crossings were riveted. Her main cabin was "elegantly finished and furnished." The interior was done in natural hard woods.

And the equipment of the HARWOOD PALMER was in keeping. Steam steering gear, two steam wrecking pumps, two bilge pumps, and a steam capstan, all designed and built by Hyde. She carried two ordinary boats and two gasoline launches.

Tempted by the high freight rate of $20 per ton for coal during World War I, her owners put her in service as a coal carrier. On May 23, 1917 the HARWOOD PALMER struck a floating mine off the French Coast. Thanks to the ample small boats, the entire crew reached shore safely.

HYDE

HYDE WINDLASS COMPANY
BATH, MAINE

East Coast Representative

Courtesy of Mr. Michael Footer

Interior of the Harwood Palmer *under construction*
Courtesy of Captains Douglas K. and Linda J. Lee

A Quiet Riverbank – Courtesy Waldoborough Historical Society

The Palmer Vessels of Waldoboro:

1. *Fannie Palmer* (named for William Palmer's daughter)
Launched November 8, 1900
266 feet X 46 X 25 feet; 2258 tons
Fate – Collision early in career with schooner *Charles D. Hall*; abandoned but later salvaged and sold under name of *George R. Hudson*. Sank July 11, 1914 – Great Round Shoal.

2. *Baker Palmer* (named for benefactor family from Boston area)
Launched October 1, 1901
284 feet X 46 X 21 feet; 2792 tons
Fate – Foundered at sea en route to Buenos Aires, Argentina – Dec. 11, 1915; crew saved.

3. *Paul Palmer* (named for William Palmer's son)
Launched August 19, 1902
276 feet X 44 X 24 feet; 2193 tons
Fate – Burned off Cape Cod – June 13, 1913

4. *Dorothy Palmer*
Launched November 28, 1903
305 feet X 48 X 27 feet; 2872 tons
Fate – Stranded at Stonehorse Shoal – March 25, 1923

5. *Singleton Palmer* (named after Woonsocket, R.I. benefactors)
Launched (in the snow!) May, 1904
269 feet X 45 X 21 feet; 2859 tons
Fate – Rammed and sunk by Clyde Liner *Apache* off Delaware coast November 6, 1921.

6. *Harwood Palmer* (named for Boston area benefactor)
Not finished by George Welt but by James Creighton, former captain of the *Fannie Palmer.*
Launched November 8, 1904
301 feet X 46 X 27 feet; 2885 tons
Fate – Sunk by mine near Nazaire, France – May 23, 1917

Closing Thoughts

The closing of the Palmer-Welt shipyard in 1904 marks the finish of our story. Primarily because of these two men – Welt with his vision and Palmer with his money – Waldoboro had been able to revive temporarily an era historian Jasper Stahl calls "the Great Days." This meant employment and prosperity similar to what there had been in the 1850's, 1860's, and 1870's – due to wooden shipbuilding and spurred on by the other smaller industries – grist milling, granite quarrying, corn canning, ice harvesting, and pottery. As the 20th century got underway, many of these staple moneymakers – and employers – of the 19th century were also being phased out because of more modern inventors. A long economic decline and recession was unfortunately moving into place and was going to remain for a while. Waldoboro's "Merchant of the Medomak" role was coming to an end.

Good Old Days? – Maybe –

In our thinking about life in Waldoboro 100 or more years ago, it is wise to avoid becoming overly sentimental. Living was slower and less complicated; there was no threat of war – or terrorism; drugs virtually unheard of; global warming and pollution not nown about so not a problem. Sounds pretty good, doesn't it, and it might have been fun; HOWEVER.... there were certain aspects that we would find very difficult – think of life without running water, household appliances, automobiles, paved streets, electricity, modern hospitals – just to name a few.

The writer has gathered a few quotes from the early pages of *The Lincoln County News* to help us all appreciate, perhaps, what we have, and what we would not want, from the "good old days". Try to imagine as best you can.

1. Without the protection of SCREENS there were FLIES everywhere!

August 2, 1877 – "Oh, the flies! The horrible flies!
Buzzing around like election lies;
Dodging about like a maniac's dream,
Over the butter and into the cream,
Holding conventions all over the bread,
Biting your ears and tickling your head.
Crawling, buzzing, too busy to die,
Doggone the nasty, pestiferous flies."

2. Before our ancestors instituted licenses and, MUCH later, leash laws, there were DOGS, uncontrolled, everywhere. (No, "pooper scoopers", either!)

September 18, 1875 – "Dog fights on the sidewalks and streets of the Village are a daily occurrence."
June 25, 1896 – "75 Village dogs have not yet been licensed. The constable has been given the right to exterminate them."

3. Homeless men were often described as TRAMPS. Treatment and opinions of these people varied:

February 2, 1896 – "Tramps are an intolerable nuisance. Three of them, when stopped by two selectmen the other night, were both impudent and threatening. Noone should feed them; if some do, they will pester others. Some towns have ordered their citizens not to feed them, and it has worked."

October 9, 1879 – "A visiting tramp spent the night in the lockup and was served food left over from The Women's Christian Temperance Union meeting held earlier upstairs. Next morning the tramp was heard to observe that when he came through Waldoboro the next time, he would be sure to invite some of his friends if such a wonderful banquet were to be served!"

Finally four inventions about to make their appearance.

4. The PHONOGRAPH

October 17, 1878 – " A phonograph was exhibited the other day in Damariscotta to a large audience. On the whole it was a disappointing experience, for although it does, in an indistinct way, reproduce a human voice and can whistle quite well, it cannot do musical instruments. It is a difficult problem for science to wrestle but, if improved, might turn out to be a valuable invention. And Mr. Edison's capabilities seem inexhaustible!"

5. The TELEPHONE

May 25, 1899 – "More telephone poles are being placed on Friendship Road. There are 35 subscribers now in the village."

6. The AUTOMOBILE

April 26, 1900 - "The first automobile (or horseless carriage) ever seen in Waldoboro passed through the village at about 11 A.M. Sunday. Mr. and Mrs. W.L. Carter left their home in Attleboro, Massachusetts, went through Portland Saturday, and were expected at their journey's end in Rockland late Sunday. Despite the very poor roads, they were able to average 15 miles per hour."

7. The AIRPLANE

May 29, 1919 – "The airship that passed over the town Monday excited a great deal of interest."

The GOOD OLD DAYS? Well, maybe!

Main Street – 1916 – Early automobiles sometimes had better success ascending steep hills backward! Author's collection

Bibliography

Books

Albion, R.G., *The Rise of New York Port 1815–1860.* Scribner's Sons, New York, 1939.

American Heritage, *Clipper Ships and Captains*, Harper and Row, 1962.

American Heritage, *The Pioneer Spirit*, Simon and Schuster, 1959.

Baker, William A., *A Maritime History of Bath, Maine and the Kennebec River*, Volumes I and II, The Marine Research Society of Bath, 1973.

Biscoe, Mark W., *No Pluckier Set of Men Anywhere*, Lincoln County Publishing Co., 1994.

Bunting, William, *A Day's Work*, Volumes I and II, Tilbury House.

Cutler, Carl C., *Greyhounds of the Sea*, New York, 1930.

– *Queens of the Western Ocean*, U.S. Naval Institute, Annapolis, Md. 1961.

Desjardins, Thomas, *Stand Firm, Ye Boys from Maine!*

Duncan, Roger F., *Coastal Maine*, W.W. Norton and Co., 1992.

Everson, Jennie G., *Tidewater Ice on the Kennebec*, Maine State Museum, 1970.

Fairburn, John, *Merchant Sail*, Volumes I – VI, Fairburn Maine Educational Foundation, Center Lovell, Maine.

Gay, Maude, *Echoes from Old Doorways.*

Goodman, John, *The California Gold Fleet,* Maine Maritime Museum.

Lubbock, Basil, *The Down Easter*, Charles Lauriat and Co. Boston, 1929.

Manning, Samuel, *The Building of a Wooden Ship.*

Marshall, Don, *Oregon Shipwrecks.*

Matthews, Frederick C., *American Merchant Ships 1850 – 1900*, Salem, Mass. Marine Research Society, 1931.

Miller, Samuel L., *Waldoboro, Maine*, Reprinted by Lincoln County Publishing Co., 1987.

Morrison, Samuel, *Oxford History of the American People*, Oxford University Press, 1965.

Parker, Lieutenant W.J. Lewis, U.S.C.G., *The Coal Schooners of New England 1870 – 1909*, Maine Historical Association Inc., 1948.

Snow, Linwood and Lee, Captain Douglas, *A Shipyard in Main; Percy and Small and the Great Schooners*, Tilbury House and Maine Maritime Museum, 1999.

Stahl, J.J., *The History of Old Broad Bay and Waldoboro*, Bond Wheelwright Co., 1956.

Whitaker, W. and Horlacher, G., *Broad Bay Pioneer*, Picton Press, 1996.

Woodward, Joseph, *Historic Record and Biographic Roster of the 21st Maine.*

Zetz, Abigail and Zoidis, Marilyn, *Woodsmen and Whigs – A History of Bangor*, The Donning Co. Bangor Historical Society, 1991.

Pamphlets, Letters, Journals

Applebee, Robert – A List of Vessels Built in Waldoboro, Maine

Baldwin, Delma – Genealogy and Vessels of Samuel Nash

Gross, Esther – "My Father's Farm"

Journal of Captain Nathaniel Groton

Letters of Edwin Clark

Letters of Albion Oakes

Letters of William Palmer

Log of Captain Matthews of the schooner *J. Manchester Haynes*

Report of the Industrial and Labor Statistics – 1889

Senior Class Essays – Waldoboro High School – 1934

Sidelinger, Lowell – A List of Vessels Built in Waldoboro, Maine

Waldoboro Business Pamphlet – 1892

Waldoboro Historical Society – "Waldoboro"

Waldoboro – Vital Records – A List of Marriages

Waldoboro – 1973 Bi-Centennial Book

Newspapers

Bath Times

The Lincoln County News

Rockland Courier-Gazette

Waldoboro Press

Boston Transcript– Daily News

Morning Oregonian

Portland Press Herald

Interviews and Talks

Philip Achorn
Stanley Bailey
J. Malcom Barter
Charles Begley
Alan Benner
William Blodgett
Martha Boggs
Louis Boissoneault
Sandra Denman
George Dow
Elsie Eugley
Marjorie Freeman
Michael Footer
John Gatcombe
Randy Gross
Stanley Hanna
George Leach
Douglas Lee
Linda Lee
Nathan Lipfert
Darryl McKinney
Delia Mohlie
Thomas Moody
John Murray, Jr.
Robert Pearson
Lewis Parker
Elizabeth Sproul
Alfred Storer
William Travers
Richard Wallace

Maps

1800 Waldoboro

1850's Maine

1853 Maine Customs Districts

1857 Lincoln County

Street Maps Waldoboro – 1884, 1891, 1896, 1904, 1927

School Districts – Waldoboro

Libraries

Bangor Public Library

Maine Maritime Museum

Penobscot Marine Museum

Waldoboro Public Library

Skidompha Library of Damariscotta

Peabody – Essex Museum

Rockland Public Library

Waldoboro Historical Society Library

Indices for *Merchant of the Medomak*

Index One – Ships, Builders, Life at Sea — Page Numbers

Index Two - Life in Waldoboro Village

Appendix 1
List of Ships built in Waldoboro 1783 – 1904

YEAR	RIG	TONNAGE	NAME	BUILDER
1783	Schr			John Ulmer et al
1786	Ship	188	*Nancy*	
1787	Brgnt	96	*Success*	
1787	Sloop	98	*Waldoborough*	
1790	Sloop	101	*Betsey*	
1790	Schr	62	*Fair Lady*	
1793	Schr	115	*Neptune*	Thomas Meguire
1793	Sloop	79	*Sally*	Charles Samson, Sr., Charles Samson, Jr., Thomas Moore
1794	Sloop	57	*Mary and Sally*	John Ulmer
1794	Sloop	84	*Mary*	
1794	Schr	31	*Elizabeth*	Andrew Hoffses, Anthony Hoffses
1795	Sloop	46	*Polly*	Joseph Ludwig, Peter Light
1798	Sloop	85	*Turkish Rover*	
1798	Sloop	106	*German Peggy*	
1798	Sloop	89	*Tryall*	
1799	Sloop	81	*Polly*	
1800	Schr	106	*Eliza*	
1801	Snow	157	*Lydia Head*	
1801	Brgnt	175	*Two Brothers*	John Anderson
1801	Schr	102	*Resolution*	
1803	Schr	119	*Trial*	
1803	Schr	110	*Independence*	
1803	Brig	150	*Moses*	Moses Carlton
1803	?	?	*Eight Sisters*	?
1806	Schr	125	*William*	David Otis, William M. Boyd
1806	Schr	103	*Rambler*	
1806	Schr	112	*Hiram*	Moses Carlton, Jr., John Woodman
1807	Schr		*Fair Lady*	
1810	Brig	175	*Charles Miller*	David Otis, William M. Boyd
1810	Schr	114	*Walter*	Nathaniel Austin, Moses Carlton, Jr.
1810	?	?	?	Merritt
1811	Sloop	95	*Joseph*	
1812	Sloop	105	*Industry*	
1812	?	?	*Bertha*	Anthony Castner
1813	Sloop	97	*Dolphin*	
1813?	Sloop	?	*Sukey*	
1813	Schr		*The New World*	Anthony Castner
1813	Schr	?	*Langammon*	Anthony Castner

1815	Brig	189	*Tom*	
1816	Schr	103	*Globe*	Anthony Castner ?
1816	Sloop	87	*Catherine*	
1816	Schr	100	*Fair Dealer*	
1816	Sloop	103	*Thomas*	
1816	Schr	116	*Olive Branch*	
1816	Schr	105	*American Eagle*	
1816	Schr	93	*Hannah and Jane*	
1816	Sloop	98	*Elizabeth*	
1817	Schr	106	*Zealous*	
1817	Schr	105	*Joseph and Mary*	
1818	Schr	101	*Fair Trader*	
1819	Schr	102	*George Washington*	
1819	Brig	174	*Francis Miller*	Charles Miller
1819	Sloop	98	*Harriet*	
1819	Schr	103	*Chas. Kaler*	
1820	Schr	99	*Fair America*	
1820	Schr	?	*Milo*	
1820	Brig	?	*Only Son*	
1820	Schr	?	*Harriet Storer*	
1820	Schr	116	*Fair Play*	
1820	?	?	*Joseph and Mary*	
1820's?	Brig	?	*Roxanne*	Joseph Miller ?
1820's?	Schr	?	*Alden Boyd*	
1820's?	Brig	?	*Pandora*	
1820's?	Schr	?	*Star*	
1823	Schr	72	*Dispatch*	
pre1823	Schr	52	*First Attempt*	
pre1823	Brig	191	*Calliope*	
pre1823	Brig	?	*Dolphin*	
pre1823	Sloop	?	*Thomas*	
pre1823	Schr	?	*Francis Maria*	
pre1823	?	?	*Enterprise*	
pre1824	Schr	131	*General Knox*	
1824	Brig	218	*Charles Miller*	Charles Miller
1824	Brig	197	*Rook*	
pre1824	?	?	*Medomak*	
1824	Schr	111	*Illuminator*	Creamer
1824	Schr	108	*Juno*	
1824	Schr	?	*Garland*	
pre1824	Sloop	?	*Champion*	
pre1824	Schr	?	*Eclipse*	
pre1824	?	?	*Susan Miller*	Charles Miller
pre1824	?	?	*William Cole*	

pre1824	Ship	?	*Adeline*	
pre1824	?	?	*Boston Packet*	
1825	Brig	?	*Betsey and Mary*	
1826	Schr	111	*Philanthropist*	
1826	Schr	106	*Trial*	
1826	Sloop	101	*Dove*	
1827	Schr	102	*Echo*	
1827	Schr	?	*Garland*	Charles Miller
1828	?	?	*Fannie Miller*	Charles Miller
1829	Schr	104	*Alden*	
1829	Schr	115	*Bayne*	Fred Castner
1829	Schr	93	*Hampton*	
1829	Schr	117	*Banner*	
1830	Schr	71	*Renown*	
1830	Schr	69	*Albert*	
1830	Schr	110	*Billow*	
1830	Schr	116	*Amaranth*	Joseph Clark
1831	Schr	114	*Angelia*	
1831	Brig	140	*Margaret*	
1831	Schr	135	*Vesta*	William Matthews
1831	Schr	107	*Atlantic*	
1831	Brig	149	*Salem*	
1831	Schr	112	*Jane Loud*	
1832	Schr	118	*Columbus* (or *Columbia)*	
1832	Schr	112	*Ocean*	J. R. Groton
1832	Schr	132	*Wilmington*	
1832	Brig	199	*Antares*	
1832	Schr	131	*Halcyon*	
1832	Schr	129	*North Carolina*	
1832	Schr	108	*Oscar*	
1832	Schr	113	*Baltic*	
1832	Schr	117	*Packet*	Samuel Nash
1832	Brig	204	*Triumph*	William Matthews
1832	Schr	89	*Free Trader*	
1833	Schr	166	*Alexander*	
1833	Schr	116	*Olive Branch*	
1833	Schr	129	*Republic*	J.R. Groton
1833	Schr	113	*Firm*	John Kaler, Jr.
1833	Schr	107	*Othello*	James Hovey
1833	Schr	133	*Example*	Fred Castner
1833	Schr	112	*Jane*	Samuel Nash
1833	Brig	175	*Grand Turk*	Joseph Clark
1833	Brig	161	*Monhegan*	John Kaler, Jr.
1833	Schr	131	*Tribune*	Reuben Miller & Co.

1833	Schr	147	*Vanda*	Samuel Nash
1833	Schr	149	*Waldoboro*	Charles Miller
1833	Schr	113	*Mora*	?
1833	Schr	90	*James*	?
1834	Brig	229	*Hockomock*	William Matthews
1834	Schr	?	*Sara Nash*	B. J. Eugley, Samuel Nash
1834	Brig	175	*Andes*	B. J. Eugley, John Kaler, Jr.
1834	Schr	109	*Bahama*	
1834	Schr	106	*Bertha*	
1834	Schr	112	*Mary Jane*	
1835	Schr	92	*Orion*	
1835	Schr	143	*Forest*	James Hovey
1835	Schr	27	*William*	
1835	Schr	119	?	James Hovey
1835	Ship	?	*Mary Ann*	Joseph Clark
1835	Brig	?	*Benjamin*	Kaler & Burkett
1835	Bark	?	*William and James*	William Matthews
1835	Brig	146	*Mentor*	William Matthews
1835	Brig	160	*Oswego*	John Lash
1836	Schr	30	*J. W. Crawford*	?
1836	Schr	125	*St. George*	Edwin Achorn
1836	Brig	218	*Lawrence Copeland*	?
1836	Brig	158	*Proxy*	?
1836	Schr	114	*Groton*	Jacob Eugley
1836	Ship	800	*Caroline Clark*	Joseph Clark
1836	Schr	113	*Medomak*	John Kaler, Jr.
1836	Schr	67	*Yankee*	Edward Benner
1836	Schr	160 ?	*Boston*	James Cook
1836	Brig	196	*Morosco*	Charles Miller
1836	Brig	118	*Tom Paine*	Kaler & Burkett
1837	Schr	113	*Baltimore*	?
1837	Brig	164	*Patriot*	?
1837	Schr	32	*Turkish Rover*	?
1837	Brig	189	*Virginia*	Reed, Haskell & Co.
1837	Schr	?	*Columbia*	Henry Kennedy
1837	Schr	111	*Jane Fish*	John Lash
1837	Bark	?	?	Reed, Haskell & Co.
1837	Schr	116	*Surplus*	J. and R. Miller
1837	Ship	480	*Avon*	Joseph Clark
1837	Brig	?	*Ceylon*	William Matthews
1837	Schr	?	?	Benner & Schwartz
1837	Schr	112	*Jane*	?
1838	Bark	349	*Elizabeth*	?
1838	Schr	148	*Oleon*	?

1838	Schr	129	*Lodi*	Fred Castner
1838	Schr	143	*Palestine*	J.R. Groton
1838	Schr	78	*Baltimore*	?
1838	Schr	?	*Moscow*	Henry Kennedy
1838	Brig	150	*St. Lawrence*	J. R. Groton
1838	Brig	200	*Antares*	James Cook
1838	Schr	38	*McDonaugh*	Sherman & Welt
1838	Brig	?	?	?
1838	Schr	117	*Watchman*	?
1838	Schr	115	*Boyne*	?
1838	Brig	170	*Mayflower*	?
1838	Schr	132	*Peru*	George Sproul
1839	Bark	304	*Ten Brothers*	J. and R. Miller
1839	Schr	130	*Edward Kent*	Samuel Nash
1839	Ship	543	*St. Andrew*	William and Alfred Storer
1839	Ship	543	*F. Andrews*	?
1839	Ship	388	*Massachusetts*	Geo. Kaler 3rd
1839	Ship	?	*Mallabar*	Joseph Clark
1839	Bark	345	*Antoleon*	?
1839	Schr	130	*Van Buren*	Samuel Nash
1839	Schr	92	*Mary Catherine*	John Lash
1839	Ship	?	?	Anchorn, Reed & Haskell
1840	Brig	250	*Georgiana*	?
1840	Bark	?	?	John Achorn & Co.
1840	Ship	?	*Gibraltar*	Shuman & Welt
1840	Schr	33	*Hudson Hewitt*	Shuman & Welt
1840	Brig	?	?	Benj. Harrison
1840	Bark	?	*Em*	Joseph Miller
1840	Schr	118	*Redondo*	Geo. Kaler 3rd
1840	Schr	140	*Lafayette*	John Lash
1840	Schr	126	*Frederick Hahn*	?
1840	Schr	104	*Armadillo*	?
1840	Brig	156	*Aldebaron*	?
1840	Schr	115	*Boyne*	?
1840	Schr	157	*Grandee*	?
1840	Schr	126	*N.F. Frothingham*	?
1840	Ship	598	*Sea Lion*	Joseph Clark
1841	Schr	128	*Mexican*	?
1841	Brig	205	*Alwiddie*	William Matthews
1841	Bark	290	*Averon*	Fred Castner
1841	Brig	260	*Washington*	J. and R. Miller
1841	Bark	276	*Toulon*	Achorn, Haskell, Reed
1841	Ship	625	*Desdemona*	Joseph Clark
1841	Schr	97	*Oliver Elizabeth*	?

1841	Bark	358	*Rammohun Roy*	?
1841	Bark	299	*St. Cloud*	Edwin Achorn
1841	Brig	279	*Toronto*	Kennedy & Welt
1841	Brig	250	*Ohio*	Samuel Nash
1841	Bark	300	*George Henry*	George Kaler 3rd
1841	Schr	?	*Waldoboro*	Benjamin Harriman
1841	Schr	309	*Avola*	Benjamin L. Harriman
1842	Ship	576	*Medemsich*	?
1842	Schr	75	?	?
1843	Bark	?	?	John Lash
1843	Bark	298	*Francis Burr*	?
1843	Schr	?	*Orland*	Welt, Reed & Co.
1843	Ship	414	*Antwerp*	James Hovey
1843	Schr	133	*Romeo*	George Kaler 3rd
1843	Ship	?	*Herman Hunroy*	William and Alfred Storer
1843	Ship	648	*Malabar*	?
1843	Brig	?	*Braganna*	George Sproul
1844	Schr	163	*Caroline Clark*	?
1844	Brig	179	*C. H. Appleton*	?
1844	Schr	137	*Montrose*	Thomas Gay
1844	Schr	169	*Sarah Nash*	Samuel Nash
1844	Schr	136	*Sarah Gardiner*	?
1844	Brig	150	*Torcello*	George Sproul
1844	Schr	164	*Pedemonte*	Reed, Welt & Co.
1844	Ship	?	*Carolus*	William and Alfred Storer
1845	Schr	165	*Civilian*	?
1845	Bark	299	*Louisiana*	Joseph Clark
1845	Schr	170	*Ontario*	?
1845	Schr	98	*Jane and Eliza*	?
1845	Schr	125	*Richmond*	George Kaler 3rd
1845	Schr	52	*Olive*	?
1845	Brig	?	*Jesso*	Genthner and Morse
1846	Schr	127	*Algoma*	Samuel Nash
1846	Schr	21	*Mary Jane*	?
1846	Brig	259	*Sarah Ann*	Samual Nash
1846	Bark	249	*William Larribee*	?
1846	Schr	?	*Oceola*	Welt, Reed & Co.
1846	Schr	599	*Sarah Ann*	Welt, Reed & Co.
1846	Schr	?	*Mary H. Chappell*	Henry Kennedy
1846	Ship	?	*Monterey*	?
1846	Brig	?	*Lani*	Henry Kennedy
1846	Schr	132	*Bay State*	Joseph Clark
1846	Brig	259	*Susannah*	Samuel Nash
1846	Schr	127	*Laguna*	Joseph Clark

1846	Schr	97	*Romeo*	Genthner and Morse
1846	1/2 Brig	158	*Elmira*	?
1846	Schr	120	*Redington*	Welt, Reed & Co.
1846	Brig	189	*Gen. Taylor*	Joseph and Reuben Miller
1846	Brig	175	*Matamoras*	?
1846	Brig	169	*J. L. Whipple*	Welt, Reed & Co.
1847	Brig	219	*Albemarle*	?
1847	Brig	174	*Markland*	George Sproul
1847	Schr	34	*Ocean*	?
1847	Schr	149	*Albano*	Samuel Nash
1847	Bark	263	*Mary Ellen*	Edwin Achorn
1847	Brig	190	*Cahoni*	Charles Vannah & Co.
1847	Bark	200	*Pedomente*	Welt & Co.
1847	Bark	299	*Chicora*	?
1847	Brig	199	*Cymbrus*	George Kaler
1847	Ship	723	*George Evans*	Joseph Clark
1847	Schr	149	*Mary Groton*	J. R. Groton
1847	Brig	175	*Markland*	George Sproul
1847	Schr	131	*Luella*	?
1847	Brig	199	*Egyptian*	George Kaler
1847	Brig	174	*Times*	J. R. Groton
1847	Schr	129	*Allegan*	J. R. Groton
1847	Schr	136	*Montrose*	*Thomas Gay*
1847	Schr	140	*Alps*	?
1848	Bark	?	*Adams*	Benjamin L. Harriman
1848	Schr	149	*Almira Ann*	?
1848	Bark	291	*Byron*	?
1848	Brig	199	*Parthenon*	?
1848	Brig	200	*Ragiline*	?
1848	Schr	?	*Albion*	Reed & Welt
1848	Schr	162	*Jenny Lind*	Samuel Nash
1848	Schr	?	*S. Gardner*	J. R. Groton
1848	Schr	?	?	William Fish
1848	Schr	?	*Canary*	Vannah & Hall
1848	Bark	385	*Amazon*	Henry Kennedy
1848	Schr	126	*Pushaw*	Genthner & Morse
1848	Schr	78	*Romeo*	?
1848	Schr	145	*Waterloo*	?
1848	Schr	105	*Juliette*	?
1848	Bark	271	*R. Adams*	B.L. Harriman
1848	Schr	141	*Alabama*	Henry Kennedy
1848	Schr	100	*Lunker Sue*	Thomas Gay
1848	Brig	199	*Lodebar*	?
1848	Brig	134	*Samuel Nash*	Samuel Nash

1848	Brig	199	*Ionian*	?
1848	Schr	148	*S.D. Hart*	Welt & Co.
1848	Ship	?	?	J. R. Groton
1848	Bark	258	*Drummond*	M.M. Rawson
1848	Brig	197	*John Dutton*	Robert Miller
1849	Bark	?	*Antelope*	Harriman & Comery
1849	Schr	?	*Orlando*	William Welt & Co.
1849	Schr	132	*Juliet*	?
1849	Bark	299	*Leo*	?
1849	Brig	163	*Angola*	?
1849	Schr	157	*Tortla*	Jacob Hahn & Co.
1849	Bark	?	*Nineveh*	J. R. Groton
1849	Ship	800	*Caroline and Mary Clark*	Joseph Clark
1849	Bark	449	*Nimrod*	J. R. Groton
1849	Schr	140	*Mohawk*	Edwin Achorn
1849	Brig	159	*Oberon*	James Hovey
1849	Schr	134	*Hudson*	Stahl & Co.
1849	Brig	148	*Susan Ludwig*	Thomas & William Achorn
1849	Schr	144	*Robert Miller*	?
1849	Schr ?	163	*Angola*	Charles Vannah & Co.
1849	Brig	174	*W. B. Crosley*	?
1849	Bark	680	*Elena*	?
1849	Schr	139	*Martha Hall*	Kennedy & Hall
1849	Schr	93	*Lady of the Ocean*	?
1849	Schr	?	*St. Marie*	J. R. Groton
1849	Bark	224	*Aerial*	?
1849	Schr	129	*Denmark*	Edwin Achorn
1849	Schr	77	*Denmark*	S. Nash & R. Miller
1850	Schr	56	*Canary*	Henry Kennedy
1850	Brig	147	*Edwin*	Joseph Clark
1850	Bark	220	*G. D. Smouse*	Samuel Nash
1850	Schr	106	*Lebannah*	?
1850 (1848 ?)	Brig	?	*Zyder Zee*	George Sproul
1850	Schr	107	*Lebrunt*	?
1850	Bark	?	*Emblem*	Edwin Achorn
1850	Schr	76	*Mary L.*	?
1850	Ship	758	*Eastern Queen*	M.M. Rawson
1850	Schr	106	*Morrill*	?
1850	Ship	699	*Muscongus*	Reed, Welt & Co.
1850	Schr	?	*Allegash*	Stahl & Co.
1850	Brig	232	*New World*	Anthony Castner ?
1850	Bark	?	*Brunette*	Thomas Gay

1850	Schr	?	*Illuminator*	Frederick Creamer
1850	Schr	99	*Cataherine Beale*	?
1850	Schr	82	*Orbit*	John A. Levansaler
1850	Bark	220	*Amelia*	?
1850	Ship	449	*Oregon*	Henry Kennedy
1850	Bark	785	*Clara*	?
1850	Schr	106	*Mariel*	Stahl & Co.
1850	Brig	145	*Loranzo*	?
1850	Ship	?	*George Evans*	Joseph Clark
1850	Bark	228	*Mary C. Porter*	?
1851	Bark	299	*Mary*	?
1851	Bark	197	*Civilian*	?
1851	Schr	103	*Henry A.*	Henry Kennedy
1851	Ship	880	*Levitt Storer*	Alfred Storer
1851	Schr	197	*Bucentaur*	Thomas Gay
1851	Ship	925	*R.B. Sumner*	Joseph Clark
1851	Bark	299	*Mary Ellen*	Edwin Achorn
1851	Brig	*239*	*William M. Groton*	James R. Groton
1851	Ship	*700*	*B.L. Harriman*	B. L. Harriman
1851	Brig	199	*Three Sisters*	J. R. Kennedy
1851	Brig	?	*Edward*	Reed, Welt & Co.
1851	Bark	299	*Mary F. Gibbs*	?
1851	Brig	563	*New England*	Young, Kennedy & Co.
1851	Brig	195	*Lucy Ann*	Henry Kennedy
1851	Bark	299	*James Cook*	James Cook
1851	Brig	231	*Julia Ann*	Schwartz & Castner
1851	Ship	?	*Coburg*	?
1851	Ship	744	*Toulion*	Henry Kennedy
1852	Ship	770	*Suratt*	?
1852	Ship	1337	*Madagascar*	?
1852	Ship	916	*Wings of the Morning*	Edwin Achorn
1852	Brig	?	*Peerless*	Samuel Nash
1852	Bark	337	*Antelope*	B. L. Harriman
1852	Ship	1360	*Edward Stanley*	Hovey, Reed, & Co.
1852	Brig	243	*Mary Francis*	?
1852	Ship	744 or 1091 ?	*Woodcock*	Edwin Achorn
1852	Schr	138	*Lucinda Jane*	?
1852	Ship	?	?	Aaron Kaler
1852	Schr	110	*Alabama*	Henry Kennedy
1853	Ship	998	*Ella Clark*	Joseph Clark
1853	Schr	?	*Moro*	Reed, Welt & Co.
1853	Bark	?	*Seabreeze*	Aaron Kaler
1853	Brig	254	*Mahala C. Comery*	?
1853	Brig	?	*D. Schwartz*	Schwartz & Comery

1853	Brig	276	*Marie Louise*	?
1853	Ship	1096	*Golden Age*	?
1853	Ship	895	*Spark of the Ocean*	W. Storer, C. Comery, E. Achorn
1853	Schr	143	*George Millard*	?
1853	Brig	?	?	William Welt & Co.
1853	Schr	?	?	Charles Vannah & Co.
1853	Schr	?	?	Reed, Welt & Co.
1853	Brig	247	*Thomas Achorn*	Thomas Achorn
1853	Bark	?	*Amazon*	Henry Kennedy
1854	Brig	258	*Henry*	Samuel Nash
1854	Ship	1200	*Alfred Storer*	Storer & Hovey
1854	Scow	98	*Rough and Ready*	Rufus
1854	Brig	200	*Shibboleth*	Joseph Clark
1854	Brig	249	*Trindelia*	Stahl & Co.
1854	Bark	396	*White Cloud*	Charles Vannah
1854	Brig	?	*Amanda Jane*	Thomas Achorn & Co.
1854	Brig	175	*Edwin*	Joseph Clark
1854	Brig	246	*E. Drummone*	Hovey, Reed & Co.
1854	Ship	1300	*E. Wilder Farley*	B. B. Haskell & Co.
1854	Ship	1269	*Ty A Byee*	?
1854	Ship	1096	*Ocean Belle*	Welt, Reed & Co.
1854	Brig	299	*Three Sisters*	?
1854	Ship	1250	*Achorn*	Edwin Achorn & Co.
1854	Ship	1242	*Chicago*	?
1854	Ship	806	*Moonlight*	Charles Vannah
1855	Ship	1148	*James Hovey*	Alfred Storer & James Hovey
1855	Bark	417	*George Allen*	Aaron Kaler
1855	Brig	269	*Leviathan*	George Caldwell & Co.
1855	Brig	270	*Ocean Wave*	Samuel Nash
1855	Brig	202	*Two Boys*	J. R. Kennedy
1855	Brig	240	*Eliza*	George Smouse
1855	Brig	197	*Susan Emily*	Young, Roberts, & Co.
1855	Brig	?	*Francis Geyer*	Samuel Nash
1855	Bark	327	*Sanford Achorn*	Thomas Achorn & Co.
1855?	Schr	86	*Algonia*	?
1855?	Schr	577	*M. B. Harriman*	B. L. Harriman
1855	Ship	820	*Moonlight*	Henry Kennedy
1855	Brig	231	*Almore*	Henry Kennedy
1856	Schr	156	*Canary*	Henry Kennedy
1856	Schr	145?	*William Jones*	Smouse & Welt
1856	Schr	199	*Damon*	Reed & Welt
1856	Ship	1700	*William F. Storer* (Later *Hamilton Fish*)	Storer & Comery
1856	Ship	1308	*Joseph Clark*	Joseph Clark

1857	Brig	338	*William C. Clark*	Caldwell, Morse, Young
1857	Brig	300	*Ambrose Light*	Alfred Storer
1857	Brig	346	*Fannie Lincoln*	Kennedy & Hall
1858	Ship	1088	*Pickforton Castle*	Charles Comery
1859	?	33	*Annie Jordan*	McIntyre, Caldwell & Co.
1859	Ship	1318	*Loch Na Gar*	Joseph Clark
1859	Schr	?	*Charles Comery*	Charles Comery
1859	Ship	?	*Village Belle*	Reed, Welt & Co.
1859	Ship	1308	*J. Webster Clark*	Joseph Clark
1859	Brig	?	*Ida C. Comery*	Schwartz & Castner
1860	Schr	231	*Sedona*	Reed, Welt & Co.
1860	Ship	1171	*Winston Merritt*	Reed, Welt & Co.
1860	Schr	257 ?	*Joseph W. Fish*	William Fish
		194 ?		
1860	Schr	199 ?	*G. W. Rawley*	Reed, Welt & Co.
		257 ?		
1860	Bark	516	*William Brown*	Joseph Clark
1860	Ship	1171	*Dharwar*	?
1860	Brig	244	*Alexander*	?
1860	Ship	1171	*Erasmus*	?
1860	Schr	239	*Frederick Fish*	William Fish
1860	Brig	244	*D.O. Castner*	Charles Comery
1861	Schr	272	*Samuel Fish*	William Fish
1861	Schr	40	*Frank Gould*	?
1861	Schr	272	*Union Flag*	William Fish
1862	Brig	325 ?	*John H. Kennedy*	Storer & Campbell
		396 ?		
1862	Ship	1240 ?	*Otis Norcross*	Joseph Clark
		1100 ?	(Later - *Gibroader Van Walig*)	
1862	Schr	177	*White Sea*	Reed, Welt & Co.
1862	Brig	378	*Ida M. Comery*	Charles Comery
				Schwartz & Castner ?
1862	Ship	1240	*B. H. Vander Worden*	?
1863	Ship	850	*Lady Emily Peel*	?
1863	Schr	350	*Stephen G. Hart*	Reed, Welt & Co.
1863	Schr	350	*S. H. Jackson*	W. F. Storer
1863	Schr	324 ?	*William Flint*	George Caldwell
		257 ?		
1863	Ship	782	*Edwin Clark*	Joseph Clark
1863	Brig	480	*Sarah F. Kennedy*	Henry Kennedy
1864	Brig	392	*Edwin H. Kennedy*	Henry Kennedy & Sons
1864	Schr	253	*Charles W. Holt*	Welt, Feyler, Reed
1864	Schr	293	*Charlotte Fish*	William Fish
1864	Ship	942	*Morning Glory*	?

1864	Ship	912	*Sarah C. Welt*	Reed, Welt & Co.
1864	Schr	276	*Josiah Whitehouse*	George Caldwell & Co.
1864	Schr	254	*Whitney Long*	William F. Storer
1864	Bark	737	*Celeste Clark*	Joseph Clark
1864?	Brig	472	*John Hastings*	Schwartz & Castner
1865	Bark	256	*Avola*	?
1865	Schr	243	*Ella Fish*	William Fish
1865	Bark	685	*Joseph W. Davis*	Reed, Welt & Co.
1865	Schr	295	*Fanny K. Shaw*	William Fish
1865	Schr	234	*Rising Sun*	Benner & Reed
1865	Schr	253	*Grace Clifton*	Benner & Reed
1865	Schr	264	*J. B. Marshall*	Alfred Storer
1865	Bark	575	*Chimbarazo*	Wilbur Newhall
1865	Schr	386	*American Eagle*	Joseph Clark
1865	Schr	310	*Hattie B.*	W. F. Storer
1865	Bark	512 ? 458 ?	*Reunion*	Charles Comery & Co.
1866	Schr	244	*Irene Meservey*	?
1866?	Bark	516	*William Brown*	Joseph Clark
1866	Bark	750	*Laura A. Kennedy*	Henry Kennedy
1866	Schr	222	?	William Fish
1866	Schr	?	?	John W. Welt & Co.
1866	Brig	348	*Mary C. Comery*	W. F. Storer
1866	Schr	244 ?	*Jesse Hart 2nd*	Reed, Welt & Co.
1866	Schr	272	*Mary W. Huppen*	Welt, Reed, and Feyler
1864	Schr	264	*Daniel Pierson*	Alfred Storer
1866?	Schr	239	*Frederick Fish*	William Fish ??
1866	Schr	275 ? 154 ?	*Starlight*	Reed, Welt & Co.
1866	Schr	267	*Alice C. Grace*	George Caldwell & Co.
1866	Schr	330	*Laura Bridgeman*	John W. Welt & Co.
1867?	Ship	1070	*Alfred Storer*	Alfred Storer ??
1867	Schr	266	*Maggie Mulvey*	Alfred Storer
1867	Schr	60	*Alice M. Gould*	?
1867?	Schr	152	*Charles Comery*	Charles Comery ??
1867	Schr	279	*Thomas Fish*	William Fish
1867	Ship	1258	*Gold Hunter*	Joseph Clark
1867	Brig	281	*Catawba*	George Caldwell
1867	Schr	267	*Agnes I. Grace*	William Fish
1867	Schr	165	*William Penn*	William Fish
1867	Bark	248	*Elodia A. Kennedy*	Henry Kennedy
1867	Bark	595	*Mary G. Reed*	Reed, Welt & Co.
1867	Bark	611	*Rosetta McNeil*	A. R. Reed & Co.
1867	Schr	111	*Nellie Belle*	Alfred Storer

1868	Schr	?	?	William Fish
1868	Schr	272	*Oneida*	Caldwell, Flanders & Co.
1868	Schr	234	*Zet Psi*	Henry Kennedy
1868	Schr	237	*Oriola*	Alfred Storer
1868	Ship	1496	*Annie Fish*	Reed, Welt & Co.
1868	Brig	233	*Isaac Howland*	Joseph Clark
1869	Brig	316	*Loretta Fish*	William Fish
1869	Ship	1122	*Alexander McNeil*	Reed, Caldwell & Co.
1869	Schr ?	185? 227 ?	*Joseph Fish*	William Fish ?
1870	Brig	463	*Annie R. Stover*	Storer & Benner
1870	Brig	336	*Joseph Clark*	Joseph Clark
1871	Schr	116	*General Hall*	Charles Comery et al
1871	Schr	210	*Samuel Hart*	Reed, Welt & Co.
1871	Schr	138	*James Hall*	Edwin Achorn
1871	Schr	339	*Edward Emerson*	Alfred Storer
1871	Schr	225	*Stephen G. Hart*	Reed, Welt & Co.
1871	Schr	58	*Boston*	?
1872	Schr	12	*Glendale*	Boyd S. Creamer
1872	Schr	287	*Nathan A. Farwell*	Reed, Caldwell & Co.
1872	Bark	923 ? 782 ?	*Nina Sheldon*	Joseph Clark
1872	Schr	175	*Georgia D. Loud*	Edwin Achorn & Son
1872	Schr	280	*Clara G. Loud*	Reed, Welt & Co.
1873	Bknt	484	*Mignon*	Joseph Clark & Son
1873	Schr	376	*Agnes E. Grace*	William Fish
1873	Schr	476	*Lula*	Alfred Storer
1873	Schr	250	*Lena R. Stover*	George W. Caldwell & Co.
1873	Schr	274	*Tannhauser*	Edwin Achorn & Son
1873	Brig	424	*Emily T. Sheldon*	Joseph Clark & Son
1873	Schr	449	*Ella M. Storer*	Kennedy & Comery
1873	Bark	873	*Alice Reed*	A. R. Reed
1873	Schr	498	*John Proctor*	A. Storer & Son
1874	Schr	377	*George E. Young*	A. Storer & Son
1874	Scow/ Schr	87	*Achorn*	Edwin Achorn & Son
1874	Schr	246	*Winnie Lawry*	George W. Caldwell & Co.
1874	Bknt	598	*Josephine*	Joseph Clark & Son
1874	Ship	1435	*Rosie Welt*	Reed, Welt & Co.
1874	Schr	435	*Mary J. Cook*	H. Kennedy & Co.
1874	Bknt	469	*Fred Eugene*	William Fish
1874	Schr	470 ? 443 ?	*Eva C. Yates*	A. R. Reed
1874	Ship	1326	*Carrie Clark*	Joseph Clark

1875	Schr	375	*John W. Welt*	A. R. Reed & Co.
1875	Bknt	448	*Florence L. Genovar*	H. Kennedy & Co.
1875	Schr	375	*Teresa L. Keene*	William F. Storer
1875	Ship	1550	*Isaac Reed*	A. R. Reed & Co.
1876	Bark	747	*Annie Reed*	Reed, Welt & Co.
1877	Ship	1661	*Mabel Clark*	Edwin O. Clark
1877	Ship	1449	*Willie Reed*	A. R. Reed & Co.
1877	Bknt	604	*Ralph M. Haywood*	A. Storer & Son
1878	Bark	758	*Fannie L. Kennedy*	H. Kennedy & Co.
1879	Brig	373	*Stacy Clark*	Edwin O. Clark
1879	Bknt	620	*Stephen G. Hart*	Welt, Flanders, Caldwell
1880	Schr	634	*Frank Harrington*	A. Storer & Son
1880	Bark	711 ? 676?	*Matanzas*	H. Kennedy & Co.
1880	Ship	1564	*Emily Reed*	A. R. Reed & Co.
1881	Schr	425	*Fannie L. Child*	Augustus Welt & Co.
1881	Schr	676 ? 744 ?	*Woodward Abrahams*	A. Storer & Son
1881	Schr	450	*Jennie L. Hall*	William Fish
1882	Schr	12	*General Hancock*	?
1882	Schr	404	*John H. Cross*	H. Kennedy & Co.
1882	Schr	443 ? 421 ?	*George H. Ames*	A. Storer & Son
1882	Schr	496	*Nantasket*	Edwin O. Clark
1882	Schr	428	*Annie B. Hoffses*	H. Kennedy & Co.
1882	Schr	474	*Stephen G. Loud*	William Fish
1883	Schr	404	*Evie B. Hall*	Edwin O. Clark
1883	Schr	9	*Bessie*	Carter & Lilly
1883	Schr	407	*Ida Francis*	H. Kennedy & Co.
1883	Schr	488	*Maggie G. Hart*	Augustus Welt & Co.
1883	Schr	550	*Joshua Baker*	A. Storer & Son
1883	Bktn	655	*E. O. Clark*	Edwin O. Clark
1883	Bktn	639	*Rachel Emery*	H. Kennedy & Co.
1883	Schr	504	*S. G. Hart*	William Fish
1884	Ship	1745	*George Curtis*	A. R. Reed
1885	Schr	756	*J. Manchester Haynes*	A. Storer & Son
1885	Schr	645	*Maggie S. Hart*	A. R. Reed
1885	Sloop	9	*John W. Caswell*	Thomas Creamer
1886	Schr	852	*Isaiah Hart*	George L. Welt & Co.
1888	Schr	1689	*Gov. Ames*	Levitt Storer
1889	Schr	1162	*Augustus Welt*	Welt & Co.
1890	Sloop	8	*No - Name*	Thomas Creamer
1890	Schr	1064	*James W. Fitch*	Levitt Storer
1891	Sloop	8	*Jennie Maud*	Thomas Creamer

1891	Schr	1220	*Hattie B. Simpson*	Reed & Co.
1892	Schr	777	*Ida C. Southard*	A.R. Reed & Co.
1892	Sloop	8	*Gracie*	Thomas Creamer
1892	Schr	751	*Madalene Cooney*	A.R. Reed & Co.
1893	Sloop	8	*City of Everett*	Thomas Creamer
1894	Sloop	5	*Ethel*	Thomas Creamer
1896	Schr	13	*Marguerite*	Thomas Creamer
1897	Schr	13	*Olive May*	Thomas Creamer
1898	Sloop	7	*Au Revoir*	Charles Carter
1899	Sloop	10	*Geraldine*	Thomas Creamer
1900	Sloop	6	*Clyde and Astor*	C. Winchenbach
1900	Schr	2075	*Fannie Palmer*	George L. Welt
1900	Schr	14	*Eliza A. Benner*	Thomas Creamer
1901	Sloop	7	*Minnie*	C. Winchenbach
1901	Schr	2240	*Baker Palmer*	George L. Welt
1902	Schr	1763	*Paul Palmer*	George L. Welt
1903	Schr	2315	*Dorothy Palmer*	George L. Welt
1904	Schr	2357	*Singleton Palmer*	George L. Welt
1904	Schr	2400	*Harwood Palmer*	George L. Welt

Appendix 2
Palmer vs. Welt
September 1903
Letters to Rockland *Courier-Gazette*

William F. Palmer

George L. Welt

Shipbuilder Welt's Statement

The Busy Waldoboro Contractor Denies That He Is Under Financial Embarrassment

The Busy Waldoboro Contractor Denies That He Is Under Financial Embarrassment.

THE reported failure of George L. Welt, the Waldoboro shipbuilder, created a great amount of interest in this section of the state, where Mr. Welt and the products of his yard are so well known. Mr. Welt's denial, as published briefly in The Courier-Gazette of last Tuesday, brought a sense of relief to many friends who had hoped that if there was any financial embarrassment the newspaper reports had exaggerated it.

For this purpose of learning Mr. Welt's side of the case, a reporter of The Courier-Gazette visited Waldoboro Tuesday afternoon. Operations in the shipyard were at full blast, and there was no outword indication at least that Mr. Welt is laboring under any unusual business difficulty. Mr. Welt, who was supervising matters, received the reporter with much cordiality, and in the privacy of his office explained the situation as viewed from his own standpoint. He does not deny that he has some liabilities arising from the construction of the schooner Dorothy Palmer, but asserts that every bill incurred by the work on the new schooner, Singleton Palmer, is being promptly settled and that the men are being paid every Saturday night. He has made no assignment and intends to make none unless actually forced to do so. He is perfectly solvent, he claims, and after the two schooners now under contract are built he will have met every obligation and will have realized a satisfactory profit.

The statement of the Waldoboro correspondent which appeared in The Courier-Gazette of Sept. 19, and to which reference is made, was as follows:

> At a meeting of the creditors of George L. Welt, held here last Saturday, no accurate statement was submitted. Estimated liabilities $60,000, assets nominal. Taking the contracts for building vessels too low is the general cause assigned for the failure. The schooner Singleton Palmer," now on the stocks, partially built, will be completed by the American Surety Co. of New York, Welt's bondsmen, under the oversight and management of Capt. Creighton of Thomaston who will command her. The loss of wages to the many workmen, some of whom have not been settled with since the first vessel was built, is the peculiarly distressing feature of this affair.

Mr. Welt's statement in detail as made to The Courier-Gazette reporter Tuesday afternoon and signed by Mr. Welt, is as follows:

"The statement made in Waldoboro correspondence of The Courier-Gazette, Sept. 19, was a damned lie, with one exception. There was a meeting of the small creditors of Waldoboro and vicinity, who assembled at the office of W. H. Miller, attorney, but I did not consider that it amounted to anything. Its purpose was to drive me into insolvency, which I hoped they would do. As for the statement, made in this article, that the American Surety Co.—my bondsmen—were completing the vessel, that also is a lie. As far as Capt. Creighton is concerned, he is here in his capacity as master to see that the contract is carried out fully—a very usual proceeding.

my reputation where I am not too well known for the article to have such effect.

"It has come to my knowledge that a member of a shipbuilding firm in another town has criticised me for taking vessel contracts at too small a figure. It might be interesting for the public to know that I have realized more, by several thousand dollars, on each of the last three vessels which I have built than his firm did in the construction of a schooner of similar size.

"Mr. Palmer is thoroughly familiar with the situation here, and confident that everything is progressing satisfactorily. At my own suggestion, the amounts which he sends from time to time, to meet the weekly payments, are turned over to Capt. Creighton for disbursement. I do this principally for the reason that I do not wish to place money where it may be devoured by land sharks. If I am permitted to carry on my own business, without interference from those who have comparatively little at stake, the sharks will receive their pay in due season."

[Signed] Geo. L. Welt.
Builder of sch. Singleton Palmer.

Elisha C. Vinal of Bath, who has been boss blacksmith at the Welt yard the past year, was present during a portion of the interview, and expressed approval of the sentiments it contained. "I have worked for Mr. Welt just an even year," said Mr. Vinal to The Courier-Gazette reporter, "and I want to say that I was never in the employ of a squarer man. I get my pay every Saturday night just as regularly as Saturday night comes. It is my opinion, based on what I have seen, that if people would let Mr. Welt alone everything would come out all right in the end."

The Welt shipyard was established by Reed, Welt & Co. over 60 years ago. George L. Welt was associated from time to time in the vessel building with his uncle, the late Augustus Welt, and later with A. R. Reed, the owner of the yard. He became associated with Augustus Welt in 1881 and built three vessels, after which he entered partnership with A. R. Reed and others and built four vessels. The last was completed in 1892 and with it ended Waldoboro's era of shipbuilding prosperity for a period of eight years. Grass eventually sprang up as a memorial to Waldoboro's slumbering industry and in the fall was laid low by the mowing machine—a scanty crop when compared with the more substantial product of busy ship carpenters. The metamorphosis, although not wholly new on the Maine coast during this period, was so striking that a Boston newspaper published a long article concerning it, and Waldoboro fell into the class of shipyard "has beens."

In the year 1900 it was announced that George L. Welt was to reopen the yard, and that large contracts were coming Waldoboro's way. The prophecy proved no idle one. First there was constructed the mammoth five-masted schooner Fannie Palmer. Then followed the Baker Palmer, the Paul Palmer and the Dorothy Palmer, all vessels of immense size and faultless construction. The dull season became eliminated as a feature of this industry in Waldoboro

September 26, 1903 – Rockland Courier-Gazette

coming event:

"In another column our readers will find the program of the anniversaries at Rockland, Oct. 7, and 8. Timely information is thus furnished with reference to the meetings. The committee of arrangements have faithfully attended to their duties. Some of their purposes have not been fulfilled, but when compelled to make some changes in the program the members of the committee have not yielded their high purpose to make the Rockland meetings worthy of the great Baptist brotherhood in the State of Maine. It is difficult to see how a more attractive program could have been arranged. All of the speakers from abroad are men who are held in deserved honor throughout our denomination. Dr. Ashmore comes to us fresh from his last period of service in China. He was last with us when the meetings were held at Bar Harbor in 1889, and no one who heard him there has forgotten the statesmanlike address which he delivered. Much has occurred in China since that time, and Dr. Ashmore at Rockland will have eager listeners to the story he will tell. Dr. Kerr Boyce Tupper of Philadelphia is the brilliant successor to the late Dr. Geo. Dana Boardman, as pastor of the First Baptist church in the city of Brotherly Love. Honored names among our brethren in the South Dr. Tupper bears, but he will receive from the Baptists of Maine a greeting for his own sake as well for the sainted men whom his name recalls. Dr. Barnes, now of Worcester, Mass., who has been away from New England for awhile, is not a stranger among us, and always has a message, helpful and inspiring. The meetings of the Education Society are on Wednesday afternoon and evening, instead of Thursday afternoon and evening as usual, in order that Dr. Barnes may be with us. In thus giving the whole of Thursday to convention matters the committee have made an arrangement which should leave upon all a strong and abiding impression in reference to our state work."

ticle, that the American Surety Co.—my bondsmen—were completing the vessel, that also is a lie. As far, as Capt. Creighton is concerned, he is here in his capacity as master to see that the contract is carried out fully—a very usual proceeding.

"The vessel is built on contract, with payments made from time to time as the work progresses. At certain periods the builder has thousands of dollars invested before he can receive a payment. Take for instance the schooner Dorothy Palmer, the largest five-masted schooner in the world and the third largest vessel of that rig in the world. At the time of her launching I had in my yard material for the schooner Singleton Palmer and the frame for another schooner which was to be a duplicate of the Singleton Palmer, amounting in all to about $45,000, for which I had not received one dollar, as, according to contract, nothing was due me until the Singleton Palmer was in frame. Every shipbuilder knows that thousands of dollars are expended in material before he receives a payment.

"I am today building the schooner Singleton Palmer, the fifth vessel for William F. Palmer of Boston, and have another under contract, the frame for which is in the yard. The six vessels constitute 60 per cent. of his entire fleet, which is the largest coasting and coaling fleet on the Atlantic coast, and if I had the facilities, I could have all his business.

"I am now considering an offer to become superintendent and general manager, at a good salary, of a large shipbuilding plant which will be established in Maine. This offer I shall probably accept, and when this business leaves Waldoboro the grass will be higher in the yard than it was when I found it, before you will find another young man who has the sand and ability to do the same work and expend a sum of money amounting to half a million dollars.

"The people of Waldoboro will doubtless recall the former condition of the shipbuilding industry here; how small vessels, one-fifth the size of the Singleton Palmer, were built during the summers, and how the yard was nearly always idle during the winters. Then the men received their pay when a vessel was completed and, not on every Saturday night as in the case under the present management.

"I am building these vessels on a good margin, and after No. 10 (the Singleton Palmer's duplicate) is completed, I can pay all my bills and have a good-sized margin remaining as profit.

"In relation to the rumors of my business emberrassment, I want to say that I have made no assignment and have no such intention. Should I be driven to such a recourse, I would probably save thousands of dollars thereby, but I will stake my reputation as a shipbuilder of 25 years' standing against the liability of taking such advantage. The business of the yard is as active and prosperous as it has been for any period in its history, and it is my intention that every man in my employ, and every other creditor, shall receive every dollar which is honestly due him.

"It is my conviction that fully 75 per cent. of the business interests in Knox and Lincoln counties could be failed in 48 hours if the creditors were inclined to force matters.

"The whole difficulty exists in the imagination of certain persons who appear to entertain a petty grudge against me, one of them being a prominent Waldoboro attorney, who, I am positive, was instrumental in having the article published in The Courier-Gazette, reflecting upon my credit, and having a possible tendency to injure

masted schooner Fannie Palmer. Then followed the Baker Palmer, the Prisci Palmer and the Dorothy Palmer, all vessels of immense size and faultless construction. The dull season became eliminated as a feature of this industry in Waldoboro and at the present time the contracts promise continuous work for more than a year, unless—

This has meant, as stated in the interview with Mr. Welt, an expenditure of about $500,000, a considerable portion of which has been distributed among 125 workmen and promptly turned into the channels of trade at home.

The schooner Singleton Palmer, now in frame, will be another magnificent addition to the Palmer fleet. She will register about 2500 tons gross, and her principal dimensions are as follows: Length over all 310 feet; length of keel, 269 feet; breadth of beam, 45 feet; depth of hold, 27 feet. She has an oak frame, the largest Mr. Welt thinks, which he ever put in any vessel. The planking and ceiling are of hard pine. There are to be three full decks, three large houses and a house for the after engine.

There are 16 tiers of keelsons and sister keelsons, three of which measure 15x15 inches and 12 tiers which are 14x14 inches. The ceiling material is 12x14 inches.

A particular feature of the vessels built in the Welt yard is the manner in which they are strapped. The Singleton Palmer has a belt strap extending entirely around the vessel, two feet below the main rail, and composed of 8¾ inch iron. Commencing in the middle of the schooner is a pair of straps fastened to the belt straps every six feet, running at an angle of 45 degrees, and extending to the short floor heads. These straps are 4½x¾ inches, each 55 feet long, making 80 pairs on each side. The strapping and belting is about two miles. The straps running at an angle are fastened at every timber crossing with an inch bolt, riveted on the ceiling, wherever they cross each other they are not riveted. This is what Mr. Palmer calls the semi-composite type. It is put on before the vessel is planked and is flush with the timber.

The schooner will have five Oregon masts 118 feet in length, the topmasts being 60 feet long. The other spars are also Oregon pine, the jib-boom and spanker each being 80 feet and the others about 50 feet each. She will spread about 10,000 yards of canvass. There will be two Baldt stockless anchors weighing 9000 pounds each, and made in Pennsylvania, and 120 fathoms of bar-link B. B. tested chain, 2¾ inches in size.

The equipment will be composed of Hyde machinery, including two engines, one forward and one aft, the whole outfit being of a modern character. The launching is expected to take place early in the spring and her place on the stocks will be taken by the vessel which is to be her duplicate.

The Singleton Palmer will be commanded by Capt. James A. Creighton of Thomaston, who was master of the schooner Fannie Palmer. He is known far and wide in marine circles as one of the smartest sea captains hailing from a town which has always been famous for them.

COLL OF D.K & L.J. LEE

A Reply to Mr. Welt.

William H. Miller, Attorney, Pays His Respects to the Shipbuilder's Statement.

Referring to the article in The Courier-Gazette of Sept. 26, entitled "Shipbuilder Welt's Statement," I have no hesitancy in acknowledging the authorship of the item in your issue of 19th instant, not on account of any ill-feeling against Mr. Welt, as I never have been on any other than the most friendly terms with him since boyhood, but I inserted that item simply because, in cases of failure in business, it is the usual and withal eminently proper thing to do. Moreover, as I am attorney for creditors of Mr. Welt, holding claims against him amounting to nearly $20,000, and besides have one action-at-law now pending against him and Sarah C. Welt, jointly, it must appear that I have "something at stake."

Mr. Welt states that there was a meeting of the "small creditors of Waldoboro and vicinity, at my office, etc." Small creditors, forsooth!

At that meeting were represented:

Cooney, Eckstein & Co.	$16,000
Hyde Windlass Co.	13,000
Damariscotta National Bank	5,000
Levitt Storer	550
J. T. Gay & Son	500
W. S. Kaler	503
A. N. Sprague	300
E. K. Eugley	127
Edwards & Walker	170
Total	$36,150

In addition to the above there were several creditors among which was the American Surety Co., which William F. Palmer represented, together with the Hyde Windlass Co. of Bath. Accepting as authentic the statements as to the liablity of this Surety Co., it is conservative to estimate that, at that meeting at least $45,000 in debts was represented. If these are the small creditors, where oh where are the large ones? If Mr. Welt's designation is correct then my statement of $60,000 liabilities is a low estimate. Doubtless $80,000 would not pay all of Mr. Welt's indebtedness, and we can furnish quite a schedule. We hoped that J. E. Moore, Esq., his attorney, who was present, would furnish us with a list, but it was not forthcoming. And here let me say that this meeting was of the creditors of George L. and Sarah C. Welt, as the latter is a signer on the paper of the former for more than twice that she was ever worth at any one time. And in this connection it appears that Sarah C. Welt on July 17, last, conveyed away to her niece whatever property she owned, and if this transfer is sustained by the courts it will leave the joint creditors also without anything behind their claims.

SARAH C. WELT - COUSIN OF GEO.

As to taking the contracts for building the vessels too low, such is the opinion of such well-known shipbuilders as A. R. Reed and Levitt Storer, than whom there are no better judges of what a vessel ought to cost.

Now, in regard to my statement that the American Surety Co. (I think that is the name of the company as Mr. Palmer gave it) would complete the building of the schooner Singleton Palmer, our authority is William F. Palmer, who addressed the meeting at length, and among other things, said: (to quote his own language) "All monies would be hereafter disbursed by Capt. James A. Creighton, who would be in Waldoboro regularly to attend to this part of the business until the Singleton shall have been finished; and that he (Palmer) should look to the Surety Co. (Mr. Welt's bondsmen) to see that the contract was fulfilled, and that company was really the other party to the contract from now out; and so far as any responsibility for completing this schooner or any supervision of the work in the shipyard, or paying out a cent of money, devolving upon Mr. Welt, he (Welt) might just as well go a-fishing from now until the schooner was launched and ready for sea!" Now I ask, what other inference could a listener draw from such remarks?

Singleton Palmer →

Mr. Palmer made further remarks derogatory to Mr. Welt's reputation for truthfulness and reliability, which, however we do not choose to repeat, as all the creditors present refrained from any personalities—with the exception of Mr. Palmer.

Concerning Mr. Welt's assets, what are they? The creditors cannot find any. If they could, some of Mr. Welt's workmen, ("land sharks" Mr. Welt terms these men) holding unpaid claims for labor on the vessels prior to the present one ranging from $850 downwards, would at once sue and attach anything in sight. Mr. Welt did, however, have an interest in certain buildings, etc. in the yard, which he conveyed to A. R. Reed a few weeks since, (consideration $2500) to prevent (as Mr. Welt said,) his former watchman, E. Kaler Eugley, from collecting $127 due him for services watching the schooner Dorothy Palmer, at $1.25 per night, (15 hours) through snow, sleet and rain. And this faithful old man, poor and needy, is one of the "land sharks!"

Our "strenuous townsman" makes the assertion in cold type, that if he had the facilities he could have all the Palmer business. Judging from William F. Palmer's remarks at the meeting, he would take exception to such an assumption on the part of Mr. Welt.

During the remainder of the time occupied in completing the present vessel, Mr. Welt (unless, as Wm. F. Palmer said at the meeting, the American Surety Co., to whom Mr. Welt made sworn statement on his present bond that he was worth $24,000, conclude to "deal with him") will be found at his place of business sending out reports that he is perfectly solvent. How much, let me ask does Mr. Welt actually own in the frame of No. 10? Now as to Mr. Welt's offer which he is considering as superintendent in a shipbuilding plant at a good salary, etc. we hope he will secure the position, and that his "sand and ability" will succeed. But it occurs to us that a similar arrangement was proposed a few months ago by this same man of "sand and ability" in our neighboring town of Damariscotta—and the citizens of that town became deeply interested to have Mr. Welt locate there and build vessels, and the tension was kept up until after Mr. Welt secured a cash loan of something like $5000 from the Damariscotta Bank, and then the bubble collapsed, the purpose for which the boom was started was accomplished, and the committee chosen to wait upon Mr. Welt at his home in Waldoboro lost him. No—the bank lost the $5000! and all they have to show for it is a joint note signed by George L. and Sarah C. Welt.

← HARWOOD PALMER

However, if there is any seriousness in this offer and it materializes, Mr. Welt will doubtless have his salary assigned ahead to keep it away from the "land sharks."

And by way of adding force to his article Mr. Welt submits the statement of his boss blacksmith Mr. Varney (not Vinal). Very well, where there is one employe in Mr. Varney's luck there are a score who have not fared so well, and we question somewhat whether Mr. Varney has had the opportunity to judge of Mr. Welt's finances as well as William F. Palmer has.

If we may rely upon [illegible] open declarations, it is very doubtful whether the frame now lying in the yard will ever be set up in Waldoboro, unless, as Mr. Palmer stated at the meeting, he can make a contract with Levitt Storer or some other reliable man to build her.

In conclusion allow me to say that we have studiously refrained from personalities, that although we may be a more inveterate liar than this "tin god on wheels" (to use a homely expression) nevertheless if he wants to pose with us in a newspaper contest as to which one of us is the biggest liar that ever struck Waldoboro, we are willing to enter the race and acknowledge ourselves beaten at the outset. If there is anybody who can match George L. Welt in that respect they will have to be an excellent vote-getter. At least, I think that Mr. Welt would lead off with a rousing vote from William F. Palmer and all of the creditors.

If this agony is to be prolonged I may have something to say hereafter concerning Mr. Welt's dealings in a fiduciary capacity, however it is immaterial to us. Let the good work go on.

William H. Miller.

Waldoboro, Sept. 28th, 1903.

Courtesy of Captains Douglas K. and Linda J. Lee

Biography of the Writer

Mark Wyman Biscoe was born in Boston, Massachusetts on August 10, 1936 to Professor and Mrs. Jonathan (Martha Richardson) Biscoe. Mark received his early education in the schools of Nyack, New York, Amherst, Massachusetts, and Orono, Maine, and graduated from Orono High School in 1953.

He entered the University of Maine in the fall of 1953 and received a Bachelor of Arts Degree in English in June of 1957. In June of 1967 he earned a Masters Degree in History from the University of Maine. From September of 1957 to September of 1963 he was a member of the Maine Air National Guard.

From 1958 until 1995 he served as a teacher of Latin, history, and athletics at the Fenn School of Concord, Massachusetts as well as being a dormitory master, Director of Athletics, and Assistant Headmaster at various times. Retiring in 1995, he now lives in Waldoboro, Maine with his wife Jane (the former Jane Elizabeth Quimby of Corinna and Farmington, Maine). Mark and Jane have three children: Mark, Andrew, and Katherine, as well as seven grandchildren: Timothy, Daniel, Elizabeth, Candace, Matthew, Emma, and Jonathan.

Merchant of the Medomak is Mark's second book. In 1994 he published *No Pluckier Set of Men Anywhere*, the story of ships and shipbuilding in Damariscotta and Newcastle, Maine.